# Seminar on Nonlinear Partial Differential Equations

Edited by S.S. Chern

With 35 Illustrations

Springer-Verlag
New York Berlin Heidelberg Tokyo

S.S. Chern
Department of Mathematics
University of California
Berkeley, CA 94720
U.S.A.

Mathematical Sciences Research Institute
2223 Fulton Street, Room 603
Berkeley, CA 94720
U.S.A.

AMS Subject Classification: 35-02, 35AXX, 35BXX, 35QXX, 73C50, 76-XX

Library of Congress Cataloging in Publication Data
Main entry under title:
Seminar on nonlinear partial differential equations.
(Mathematical Sciences Research Institute
publications; 2)
Includes bibliographies.
1. Differential equations, Partial–Addresses, essays,
lectures. 2. Differential equations, Nonlinear–Addresses, essays, lectures. I. Chern, Chiing–Shen.
II. Series.
QA377.S45 1984 515.3'53 84-20210

The Mathematical Sciences Research Institute wishes to acknowledge support from the National Science Foundation.



Printed and bound by R.R. Donnelley & Sons, Harrisonburg, Virginia.
Printed in the United States of America.



9 8 7 6 5 4 3 2 1

ISBN 0-387-96079-1 Springer-Verlag New York Berlin Heidelberg Tokyo
ISBN 3-540-96079-1 Springer-Verlag Berlin Heidelberg New York Tokyo

# Foreword

When the Mathematical Sciences Research Institute was started in the Fall of 1982, one of the programs was "non-linear partial differential equations". A seminar was organized whose audience consisted of graduate students of the University and mature mathematicians who are not experts in the field. This volume contains 18 of these lectures. An effort is made to have an adequate Bibliography for further information. The Editor wishes to take this opportunity to thank all the speakers and the authors of the articles presented in this volume for their cooperation.

S. S. Chern, Editor

## Table of Contents

# GEOMETRICAL AND ANALYTICAL QUESTIONS IN NONLINEAR ELASTICITY

by

Stuart S. Antman

Department of Mathematics
and Institute for Physical Science and Technology
University of Maryland
College Park, MD 20742

## 1. Introduction

There are many reasons why nonlinear elasticity is not widely known in the scientific community: (i) It is basically a new science whose mathematical structure is only now becoming clear. (ii) Reliable expositions of the theory often take a couple of hundred pages to get to the heart of the matter. (iii) Many expositions are written in a complicated indicial notation that boggles the eye and turns the stomach.

One may think of nonlinear elasticity as characterized by a quasilinear system of partial differential equations. But there is more to the subject than the mere analysis of these equations, however difficult it may be: These equations are augmented by a variety of subsidiary conditions reflecting that nonlinear elasticity is meant to describe the deformation of three-dimensional bodies in accord with certain laws of physics. Thus the subject includes all the richness and complexity of three-dimensional differential geometry.

In this article, I give a concise presentation of the basic theory of nonlinear elasticity. The mathematical structure of the equations is examined with emphasis placed on the inextricable linking of geometry, analysis, and mechanics. The article ends with a comparison of nonlinear elasticity with the much better known theory of incompressible Newtonian fluids.

To remove some of the mystery from nonlinear elasticity, we give a very brief sketch of how the conceptual foundations lead to the governing equations. The rest of this article is devoted to an elaboration of these ideas, which will show why the subject is rich, fascinating, and challenging.

We identify material points of a body by their positions x in some reference configuration. Let p(x,t) be the position occupied by x at time t. p may be thought of as the basic unknown of the problem. Its derivative $\frac{\partial p}{\partial x}(x,t)$ at (x,t) describes the local way the material is deformed and rotated. Force intensity per unit reference area at (x,t) is measured by the Piola-Kirchhoff stress-tensor T(x,t). The requirement that the force on each part of the body equal the time rate of change of the linear momentum leads the equations of motion

$$\text{(1.1)} \qquad \operatorname{Div} T + f = \rho \frac{\partial^2 p}{\partial t^2},$$

where f is a given force intensity (such as gravity) applied to the body and $\rho$ is a mass density. The form of the divergence Div T is discussed in Section 3. The mechanical behavior of a material is specified by a relation between $\partial p/\partial x$ and T. A material is <u>elastic</u> if there is a function $\hat{T}$ such that

$$\text{(1.2)} \qquad T(x,t) = \hat{T}\left[\frac{\partial p}{\partial x}(x,t),\ x\right].$$

The substitution of (1.2) into (1.1) yields a quasilinear system of partial differential equations for p. This system is the object of our study. We now examine with care each of the three central concepts used to systematize the theory of continuum mechanics, namely, the geometry of deformation, stress and the equations of motion, and the specification of material properties by means of $\hat{T}$. Note that the form of $\hat{T}$ completely determines the analytical properties of the system (1.1), (1.2) since it is the only thing that can be appreciably varied.

We usually denote scalars by lower case Greek letters, vectors (i.e., elements of Euclidean 3-space $\mathbf{E}^3$) by lower case Latin letters, and tensors (i.e., linear transformations from $\mathbf{E}^3$ to itself) by upper case Latin letters.

## 2. Geometry of Deformation.

In this section we study the deformation of material bodies in Euclidean 3-space $\mathbb{E}^3$, emphasizing those features most important for the mechanics of solids. It is useful to exercise some care in the definitions so that primitive physical concepts can be given precise mathematical interpretations.

A (material) body is a set, whose elements are called material points, that can occupy regions of $\mathbb{E}^3$, i.e., that can be mapped homeomorphically onto regions of $\mathbb{E}^3$. Each such mapping is a configuration of the body. We select one such configuration, called the reference configuration, in some convenient way. (E.g., it could be the configuration actually occupied by the body at some initial time or it could be some natural configuration in which each part of the body is free of force.) We can thus identify the material points of a body with the positions x they occupy in the reference configuration. For simplicity we assume that the region occupied in this configuration is the closure cl $\Omega$ of a domain $\Omega$ of $\mathbb{E}^3$. We call cl $\Omega$ itself the body.

Let p(x,t) be the position of material point x at time t. The domain of p is $(\mathrm{cl}\ \Omega) \times \mathbb{R}$. We require that $p(\cdot,t)$ be one-to-one so that two distinct material points cannot simultaneously occupy the same position in space. This injectivity is a global restriction on p, which we intend to restrict locally by requiring it to satisfy differential equations. A satisfactory treatment of injectivity, which would among other things require a careful specification of boundary conditions to hold when two distinct parts of the boundary $\partial\Omega$ of $\Omega$ come into contact, is beyond our present mathematical resources. We consequently content ourselves with requiring that the deformation $p(\cdot,t)$ preserve orientation (at least almost everywhere). In particular, if $B_r(x)$ is the open ball of radius r about material point x, then we require that

$$\lim_{r\to 0} \frac{\mathrm{vol}(p(B_r(x) \cap \Omega, t))}{\mathrm{vol}(B_r(x) \cap \Omega)} > 0 \tag{2.1}$$

for almost all (x,t) in $\mathrm{cl}\Omega \times \mathbb{R}$.

We have not required that $p(\cdot,t)$ be continuous, because the body $\Omega$ could break. (The study of fracture is a highly developed discipline in solid mechanics.) Suppose for the time being that $p(\cdot,t)$ is differentiable and denote its derivative by $\partial p/\partial x$. Then we can strengthen (2.1) by requiring that

$$\det\left[\frac{\partial p}{\partial x}(x,t)\right] > 0 \tag{2.2}$$

for each $(x,t)$ in $(\mathrm{cl}\ \Omega) \times \mathbb{R}$. In practice, we do not seek solutions p of (1.1), (1.2) among continuously differentiable functions but in larger spaces. For static problems in which p is independent of t, a natural class of spaces are the Sobolev spaces $W^1_\alpha(\Omega)$, $\alpha \in [1,\infty)$ consisting of all distributions p that together with their distributional derivatives $\partial p/\partial x$ belong to $L_\alpha(\Omega)$:

$$\int_\Omega \left[ |p(x)|^\alpha + \left|\frac{\partial p}{\partial x}(x)\right|^\alpha \right] dv(x) < \infty. \tag{2.3}$$

Here dv denotes the differential volume. For such functions it is not immediately evident how to interpret (2.2). Ball (1977) has resolved this question, using the fact that the Jacobian can be written as a divergence.

Except for regularity questions, some version of (2.2) defines the class of deformations p among which we shall seek solutions of the governing equations. We note, however, that the set of p's (in $C^1(\mathrm{cl}\Omega)$, say) satisfying (2.2) is not convex. (It may even have a countable infinity of nonconvex components.) This fact portends difficulties for the analysis.

Since (2.2) holds, the polar decomposition theorem says that $\partial p/\partial x$ can be uniquely written in the form

$$\partial p/\partial x = RC^{1/2} \tag{2.4}$$

where R is the proper orthogonal rotation tensor and $C^{1/2}$ denotes the positive-definite square root of the Cauchy-Green deformation tensor

(2.5) $$C = (\partial p/\partial x)^*(\partial p/\partial x).$$

Here the asterisk denotes transpose. C accounts for the local change of shape.

We note in passing that if C is given in $C^2(\mathrm{cl}\ \Omega)$ and if $\Omega$ is simply-connected, then the necessary and sufficient condition for (2.5) to have a local solution p, unique to within a rigid body motion, is that the Riemann-Christoffel curvature tensor based on C vanish. I know of no such result when the smoothness of C is appreciably reduced, say by requiring $C^{1/2} \in L_\alpha(\Omega)$ (corresponding to $p \in W_\alpha^1(\Omega)$).

## 3. Stress and the Equations of Motion.

The intensity of force per unit area has a much greater effect on changing the shape of a body than does the total force itself. (Hence we cut meat with a knife and not with a thumb.) Stress is a tensor (linear transformation) characterizing this intensity. There are several kinds of stresses used in continuum mechanics. Here we introduce the one most useful in solid mechanics.

Let B be a body with $B \subset \Omega$ and with $\partial B$ sufficiently smooth (say that $\partial B$ has a locally Lipschitz continuous graph). In the configuration occupied by B at time t we assume that the resultant force on B is (the vector)

$$(3.1) \qquad \int_{B} f(x,t)dv(x) + \int_{\partial B} s(x,t)da(x).$$

Note that the integrations are taken over the body B and its boundary $\partial B$, not over the region $p(B,t)$ occupied by B at time t and its boundary $\partial p(B,t)$. We assume that $f(\cdot,t)$ and $s(\cdot,t)$ are respectively integrable over B and $\partial B$. The vector $f(x,t)$ is the force intensity per unit <u>reference</u> volume at $(x,t)$ and the vector $s(x,t)$ is the force intensity per unit <u>reference</u> area of $\partial B$ at $(x,t)$. We may allow f to depend on p. We discuss boundary values of s in Section 4.

Let $\rho(x)$ be the mass density per unit <u>reference</u> volume at x. It is a given function. The requirement that the resultant force on B at time t equal the time rate of change of linear momentum of B at t yields

$$(3.2) \qquad \int_{B} f(x,t)dv(x) + \int_{\partial B} s(x,t)da(x) = \frac{d}{dt}\int_{B} \rho(x)\frac{\partial p}{\partial t}(x,t)dv(x).$$

(This general version of Newton's Second Law is due to Euler.) We can replace (3.2) by the more general linear impulse — momentum law

$$(3.3)\quad \int_{t_1}^{t_2} \left[ \int_B f(x,t)dv(x) + \int_{\partial B} s(x,t)da(x) \right] dt$$

$$= \int_B \rho(x) \left[ \frac{\partial p}{\partial t}(x,t_2) - \frac{\partial p}{\partial t}(x,t_1) \right] dv(x).$$

This equation should hold for all $B \subset \Omega$ and for all $(t_1,t_2)$ for which the integrals of (3.3) make sense.

In a similar way the requirement that the resultant torque (about 0) on B at time t equal the time rate of change of angular momentum of B (about 0) yields

$$(3.4)\quad \int_B p(x,t) \times f(x,t)dv(x) + \int_{\partial B} p(x,t) \times s(x,t)da(x)$$

$$= \frac{d}{dt} \int_B \rho(x)\ p(x,t) \times \frac{\partial p}{\partial t}(x,t)\ dv(x).$$

(This principle is due entirely to Euler.) Here we have assumed that there are no distributed couples, i.e., that the torque consists solely of moments of the force distribution. We may readily obtain the corresponding angular impulse — momentum law, analogous to (3.3).

In order to expedite our analysis, we assume for the time being that every function (and derivative) we exhibit is continuous. Now $s(x,t)$ appearing in (3.2) actually depends on $\partial B$. If we give careful thought to the nature of contact forces between bodies, we are led to the assumption that $s(x,t)$ depends upon $\partial B$ only through the unit outer normal $n(x)$ to $\partial B$ at x. (We comment on this assumption below. If $\partial B$ has a locally Lipschitz continuous graph, then it has a normal a.e.) Note that n is normal to the material surface $\partial B$ and not to the geometric surface $p(\partial B,t)$ occupied by $\partial B$ at time t. Let us therefore set $s(x,t) = \hat{s}(x,t,n(x))$ and substitute this expression into (3.2), which is to hold for each $B \subset \Omega$ with a sufficiently regular boundary.

Now we can follow Cauchy and take $B_\gamma$ to be a tetrahedron whose faces are formed from the coordinate planes through x and a plane with unit normal n(x), having distance $\gamma > 0$ from x. We replace B of (3.2) with $B_\gamma$, divide the resulting equation by $\gamma^2$, and then let $\gamma \longrightarrow 0$. Since every function in (3.2) is continuous, we can use the mean value theorem to effect this limit. The terms corresponding to integrals over $B_\gamma$ vanish in the limit and we obtain that $\hat{s}$ is linear in n, i.e., there is a tensor T(x,t) such that

$$(3.5) \qquad \hat{s}(x,t,n(x)) = T(x,t)n(x).$$

T(x,t) is the <u>first Piola-Kirchhoff stress tensor</u> at (x,t). T(x,t)n(x) is the force per unit <u>reference</u> area at (x,t) exerted by the material of $\{ y \in \mathrm{cl}\Omega: (y-x)\cdot n > 0 \}$ on that of $\{ y \in \mathrm{cl}\Omega: (y-x)\cdot n \leqslant 0 \}$.

We can now substitute (3.5) into (3.2), use Green's theorem to convert the integral over $\partial B$ into one over B, and then use the continuity of the integrands and the arbitrariness of B to obtain the <u>classical form of the equations of motion</u> for a continuum:

$$(3.6) \qquad \mathrm{Div}\ T + f = \rho \partial^2 p/\partial t^2.$$

The actual form of the divergence "Div" can be determined from (3.2) by using components or by using a properly invariant form of Green's Theorem.

We similarly substitute (3.5) into (3.4) and reduce it to local form. We then observe that many terms in it vanish by virtue of (3.6). We thus obtain the symmetry condition

$$(3.7) \qquad T(\partial p/\partial x)^* = (\partial p/\partial x)T^*.$$

**Remarks.** We blithely made several assumptions in this development without the comment they deserve. Let us examine these issues.

(i) We never defined force in a mathematical sense; we simply imposed (3.1). More generally, if we restrict B appropriately, we could say that the resultant force at time t acting on a subbody B in $\Omega$ is the value at B of a vector-valued measure defined on $\mathrm{cl}\Omega$. This notion can be constructed from more primitive notions describing the force exerted by one body on another, in a theory developed by Noll. Cf. Truesdell (1977). In this setting, (3.1) says that the resultant force is a sum of two measures, one absolutely continuous with respect to Lebesgue volume measure and one absolutely continuous with respect to surface measure.

One virtue of such an approach is that it leads to a resolution of Hilbert's sixth problem for continuum mechanics. This purpose has not enlisted the enthusiasm of many workers in mechanics. Another objective is to show that T, which, as far as its appearance in (3.6) is concerned, is determined only to within a function with a vanishing divergence, is a well-defined mechanical concept reflecting a realistic, primitive notion of force. But the most compelling reason for scrutinizing the foundations is that problems for (1.1), (1.2) could admit solutions that are very irregular. When this occurs, the very hypotheses leading to various weakened forms of (1.1) and (1.2) may not be valid. Only through a careful analysis of the foundations can the physical significance of irregular solutions be ascertained.

(ii) Part of Noll's analysis described above leads from primitive hypotheses to a proof of our assumption that the dependence of s on $\partial B$ is only through the unit normal vector.

(iii) The derivation of (3.5) from the dependence of s on n has been carried out by Gurtin and Martins (1976) without the regularity assumptions used by Cauchy.

## 4. Boundary and Related Conditions.

The reader who has studied the three classical second-order linear partial differential equations is familiar with the Dirichlet and Neumann boundary conditions. We shall be confronting a system of three second-order differential equations, which admit a much richer array of physically reasonable boundary conditions.

Let x be a material point of $\partial\Omega$ at which $\partial\Omega$ has a unit outer normal vector n(x). At any fixed instant of time t, we could

0) prescribe the position of x,

1) constrain x to move on a given curve (which is allowed to move) and prescribe the component of T(x,t)n(x) tangent to the curve to be a given function (of p(x,t),x,t, say),

2) constrain x to move on a given (moving) surface and prescribe the components of T(x,t)n(x) tangent to the surface to be given functions (of p(x,t),x,t, say),

3) prescribe T(x,t)n(x) to be a given function (of p(x,t),x,t ).

The number identifying each alternative is the number of degrees of freedom that the material point $x \in \partial\Omega$ has at time t. Case (0) is the analog of the Dirichlet condition and case (3) the analog of the Neumann condition. We may also append initial conditions specifying p and $\partial p/\partial t$ over $\Omega$ at time t.

We now show that a set of mixed conditions in which conditions of type 0 hold on part of $\partial B$ and conditions of type 3 hold on the remainder yield some fascinating intricacies not obvious at first glance, intricacies reflecting the underlying geometry, which do not arise for scalar equations.

Consider a shaft whose lateral surface is stress-free and whose ends have prescribed configurations (obtained, say, by welding them to rigid walls). Now suppose that one end of the shaft is twisted about its axis through an angle that is an integral multiple of $2\pi$. The positions of the material points of the ends are the same.

Nevertheless, the configurations of the shaft are quite different: One can have zero stresses and the other large stresses. We are faced with the problem of distinguishing the two kinds of configurations with the same boundary conditions in an analytically convenient way.

Before attacking this problem head-on, it is prudent to perform some experiments to see the sort of geometrical difficulties that may be encountered. Consider a strip obtained as the intersection with the shaft of a plane through the axis of the shaft. It is clear that we can describe the twist suffered by the shaft if we can describe the twist suffered by a strip.

We can perform actual experiments with a strip made from a thick rubber band or packing tape. Suppose one end of the strip is twisted through an angle $4\pi$ relative to the other end, while the strip is held taut. The edges look helical. If the strip is allowed to go slack, then the twisted state may jump to one consisting of loops. This fact suggests caution is needed in prescribing the amount of twist. If the twisted strip is caused to cut a support (i.e., an arm holding an end) in an appropriate way, then the strip resumes its original flat state. (The cutting can be effected painlessly by releasing one end of the strip, moving the strip appropriately and then grasping it again.) Illustrations of this process are given by Alexander and Antman (1982).

We seek a functional $\tau$ of p, called a twist functional, for the shaft (or for the strip) whose specification fixes the amount of twist. As the experiments with strips indicate, the definition must be subtle because topological considerations are involved. Indeed, such a functional must not only be geometrically reasonable, but must also meet analytic and physical restrictions. Thus we require that $\tau$ have the following properties:

i) $\tau$ is a deformation invariant, so that different configurations giving a twist functional the same value can be deformed into each other. (The notion of deformation must be made precise). The class of all deformations for which $\tau$ has the same value should be easy to identify.

ii) $\tau$ has a mathematical form suitable for an analysis of boundary value problems for which it serves as a side condition. Since such boundary value problems are frequently handled by the calculus of variations, such functionals should be weakly continuous on the function space in which the problem is posed.

iii) $\tau$ should be a null Lagrangian, i.e., its Euler-Lagrange equation should be identically satisfied. If this condition were not met, then the weak form of the equilibrium equations, described in the next section, would require that the constraint embodied in the prescription of the value of $\tau$ must be maintained by constraint forces not accounted for by f.

This mechanical requirement (iii) says that the derivative of $\tau$ is zero, so that $\tau$ is constant on connected parts of its domain. Thus criterion (iii) actually embodies criteria (i) and (ii).

There are several ways to define twist functionals. They each rely on the observation that the twist suffered by a strip is determined by the deformation of the material curves forming its edges. One of the most effective twist functionals is based upon the linking number of Gauss. To use this invariant the two edge curves are each continued to form closed loops. These continuations, which are regarded as unmovable, correspond to the support holding the shaft. The linking number for the two closed loops is a twist functional satisfying criteria (i) - (iii) when the class of deformations is restricted to those for which one complete loop never slices the other. The full details of the problem of prescribing twist have been worked out by Alexander and Antman (1982).

## 5. Weak Form of the Equations of Motion.

Let us decompose $\partial\Omega$ into a disjoint union of four sets $\Gamma_0(t)$, $\Gamma_1(t)$, $\Gamma_2(t)$, $\Gamma_3(t)$ corresponding to boundary conditions of types 0,1,2,3 described in Section 4. We could also specify all these four cases of boundary conditions by a single, compact formalism (which exploits the concept of generalized coordinates for analytical dynamics, cf. Antman and Osborn (1979)). For simplicity, we suppose that $\Gamma_1(t)$ and $\Gamma_2(t)$ are empty for all t. Then the classical forms of our boundary conditions are

$$(5.1) \qquad p(x,t) = \bar{p}(x,t) \qquad \text{for } x \in \Gamma_0(t),$$

$$(5.2) \qquad T(x,t)n(x) = \bar{s}(p(x,t),x,t) \qquad \text{for } x \in \Gamma_3(t),$$

where $\bar{p}$ is prescribed on $\cup[\Gamma_0(t) \times \{t\}]$ and $\bar{s}$ is prescribed on $\mathbb{E}^3 \times \cup[\Gamma_3(t) \times \{t\}]$.

Let us also prescribe initial conditions at time 0 in the classical form:

$$(5.3) \qquad p(x,0) = p_0(x), \qquad x \in \Omega,$$

$$(5.4) \qquad \frac{\partial p}{\partial t}(x,0) = p_1(x), \qquad x \in \Omega.$$

We can now absorb (5.2) and (5.4) into the following definitive form of (3.3):

$$(5.5) \qquad \int_0^T \Bigg[ \int_B f(x,t)dv(x) + \int_{\partial B \cap \Gamma_3} \bar{s}(p(x,t),x,t)da(x)$$

$$+ \int_{\partial B \setminus \Gamma_3} T(x,t)n(x)da(x) \Bigg] dt$$

$$= \int_B \rho(x) \left[ \frac{\partial p}{\partial t}(x,\tau) - p_1(x) \right] dv(x)$$

for "almost all" B in $\Omega$ and $\tau > 0$. This expression of the impulse-momentum law makes sense under far weaker assumptions on the regularity of the functions appearing in it than we used in our formal development. (If T is integrable over $\Omega \times [0,\infty)$, then we can define the integral of Tn over almost all sets $\partial B \setminus \Gamma_3$ by means of Fubini's Theorem.)

Equation (5.5) affords a weak version of the equations of motion, but a version not used by analysts. Their form is traditionally obtained by introducing a suitably smooth test function q satisfying

$$(5.6) \qquad q(x,t) = 0 \text{ for } x \in \Gamma_0(t) \text{ and for } t \text{ large},$$

taking an inner product of (3.6) with q, integrating the resulting equation by parts, and exploiting (5.2) and (5.4) to obtain the standard weak form (known by physicists as the <u>Principle of Virtual Power</u> (or <u>Work</u>)):

$$(5.7) \qquad \int_0^\infty \left[ - \int_\Omega \underset{\sim}{T}^{*}:(\partial q/\partial x)\, dv + \int_\Omega f \cdot q\, dv + \int_{\Gamma_3(t)} \bar{s} \cdot q\, da \right] dt$$

$$= - \int_0^\infty \int_\Omega \rho \left[ \frac{\partial p}{\partial t} \cdot \frac{\partial q}{\partial t} \right] dv\, dt - \int_\Omega \underset{\sim}{p}_1(x) \cdot \underset{\sim}{q}(x,0)\, dv.$$

The colon in the first integrand is the inner product on the space of linear transformations.

There are several objections to this derivation of the weak form of the equations: (i) It employs the classical form of the equations even though the basic purpose of the weak form is to supersede the classical form precisely when the classical form is not known to be valid. (ii) Equivalent versions of the classical equation can be obtained by multiplying it by some non-vanishing function of the unknowns, but the resulting weak form is not equivalent to that we obtained. (iii) The relationship between precisely formulated versions of (5.5) and (5.7) is not obvious. Antman and Osborn (1979) have given a simple construction by which (5.5) and (5.7) can be converted directly into each other with the nakedness of each term covered with integrals at all times. This construction shows that (5.5) and (5.7) are equivalent whenever the integrals of (5.7) make sense as Lebesgue integrals.

## 6. Constitutive Equations.

As we mentioned in Section 1, the most important part of the system (1.1), (1.2) is the constitutive equation (1.2). All questions of analysis devolve on its form. We now show how its form must account for a variety of geometrical and mechanical considerations.

It should be noted that there is scant experimental information on the form of $\hat{T}$ for real materials. This fact allows mathematical analysis to guide physical investigation rather than to give it an a posteriori certification of legitimacy. In this setting we openly acknowledge our ignorance and consider whole classes of materials at one time. Our aim is to determine those differences in the functions $\hat{T}$ that lead to qualitatively different behavior of solutions. This approach is completely in the spirit of the modern theory of nonlinear operators.

We now formalize the ideas surrounding (1.2). Let $\mathbb{L}$ denote the space of all linear transformations from $\mathbb{E}^3$ to itself, let $\mathbb{L}^+$ denote those transformations having a positive determinate, and let $\mathbb{P}$ denote the symmetric, positive-definite transformations. Then the material at x in $\Omega$ is <u>elastic</u> if there is a function

$$(6.1) \qquad \mathbb{L}^+ \ni G \longmapsto \hat{T}(G,x) \in \mathbb{L}$$

such that

$$(6.2) \qquad T(x,t) = \hat{T}\left[ \frac{\partial p}{\partial x}(x,t),x \right].$$

A body is elastic if each of its material points is elastic. $\hat{T}$ does not depend on p itself, for if it did, then the material properties would change when the body is translated. To ensure that material properties are unaffected by rigid motions, we require that

$$(6.3) \qquad Q\hat{T}(G,x) = \hat{T}(QG,x)$$

for all G in $\mathbb{L}^+$ and for all proper orthogonal Q. Condition (6.3) holds if and only if there is a tensor function $\mathbb{P} \ni C \longmapsto S(C,x) \in \mathbb{L}$ (called the <u>second</u> <u>Piola-Kirchhoff</u> <u>tensor</u>) such that

(6.4a) $$\hat{T}(G,x) = GS(G^*G,x)$$

(cf. (2.5)). Thus (6.2) becomes

(6.4b) $$T(x,t) = \frac{\partial p}{\partial x}(x,t)S(C(x,t),x).$$

We require that S be symmetric

(6.5) $$S^* = S.$$

Then (3.7) is automatically satisfied and need no longer concern us. Henceforth we incorporate (6.4) and (6.5) into our definition of elasticity. It is convenient to carry along both functions $\hat{T}$ and S in our exposition. Each has special algebraic advantages for different circumstances.

The material at x is called <u>hyperelastic</u> if there are functions

(6.6) $$\mathbb{L}^+ \ni G \longmapsto \phi(G,x) \in \mathbb{R}, \qquad \mathbb{P} \ni C \longmapsto \psi(C,x) \in \mathbb{R}$$

with

(6.7) $$\phi(G,x) = \psi(G^*G,x)$$

such that

(6.8) $$\hat{T}(G,x) = \partial\phi(G,x)/\partial G$$

and equivalently

(6.9) $$S(C,x) = 2\partial\psi(C,x)/\partial C.$$

$\phi$ and $\psi$ are called <u>stored</u> <u>energy</u> <u>functions</u>. There are strong thermodynamical reasons for adopting (6.8), (6.9). If the system of loads applied to a hyperelastic body is conservative, then the system (1.1), (1.2) is equivalent to the Euler-Lagrange equations for an appropriate Lagrangian functional so that (1.1), (1.2) can be studied by variational methods.

An <u>isotropic</u> material is one having no preferred directions. It is characterized by the requirement that

$$(6.10) \qquad QS(C,x)Q^* = S(QCQ^*,x)$$

for all orthogonal Q. Materials, such as crystals, which are not isotropic, can satisfy (6.10) for Q in a proper subgroup of the orthogonal group. There is an extensive theory for the representation of S in these cases. In particular, a material is isotropic at x if and only if S has the form

$$(6.11) \quad S(C,x) = \alpha(j,x)I + \beta(j,x)C + \gamma(j,x)CC$$

where j stands for the three principal invariants of C.

We now discuss the nature of physically reasonable restrictions on $\hat{T}$. Then we shall examine their analytical consequences. Armchair experiments suggest two conclusions:

i) To effect an extreme deformation, i.e., one in which some material fiber is stretched to infinite length or compressed to zero length, an infinite amount of stress is required.

ii) The more an elastic body is pulled in one direction, the longer it grows in that direction. The more shearing force applied to a body, the more shear deformation in the same sense is produced, etc.

We wish to convert these observations into precise mathematical restrictions on $\hat{T}$. The first difficulty in doing so is caused by the imprecision of statements (i) and (ii). There are many ways to measure stress. (We have seen two, T and S, and we introduce a third stress $\Sigma$ in Section 8.) There are even more ways to measure strain, e.g., by C, $C^{1/2}$, etc.

We shall interpret (i) to mean

$$(6.12) \qquad \left| \hat{T}(G,x) \right| \longrightarrow \infty \qquad \text{as any eigenvalue of } G^*G \text{ approaches } 0 \text{ or } \infty.$$

(This condition forces $\hat{T}(\cdot,x)$ to be nonlinear.) A stronger restriction than (6.12) for hyperelastic materials is

$$(6.13) \qquad \phi(G,x) \longrightarrow \infty \qquad \text{as any eigenvalue of } G^*G \text{ approaches } 0 \text{ or } \infty.$$

The interpretation of (ii) is more delicate, not only for the reasons already mentioned, but because an elongation in one direction is accompanied by contractions in transverse directions. This coupling of effects associated with different directions is the source of the richness and difficulty in the system (1.1). (1.2). A mathematical characterization of (ii) is that the mapping from strains to stresses is <u>order-preserving</u> in some as yet unspecified sense.

The most mathematically attractive notion of order-preservation is that $\hat{T}(\cdot,x)$ be <u>strictly</u> <u>monotone</u>:

$$(6.14) \qquad \left[\hat{T}(G+H,x) - \hat{T}(G,x)\right]:H > 0, \qquad \forall\, G \in \mathbb{L}^+$$

and

$$(6.15) \qquad \forall\, H \neq 0 \text{ such that } G + \alpha H \in \mathbb{L}^+ \text{ for } \alpha \in [0,1].$$

Here ":" represents the inner product on $\mathbb{L}$. If $\hat{T}(\cdot,x)$ is differentiable, then a slightly stronger restriction is that

$$(6.16) \qquad H:\frac{\partial \hat{T}}{\partial G}(G,x):H > 0, \qquad \forall\, G \in \mathbb{L}^+$$

and

(6.17) $\forall\, H \neq 0.$

For a hyperelastic material, the analog of (6.14), (6.15) is that $\phi(\cdot,x)$ be strictly convex.

Now if $\hat{T}(\cdot,x)$ were defined on all of $\mathbb{L}$ and if it satisfied suitable growth conditions, then boundary value problems for the static version of (1.1). (1.2) have an extensive theory: Weak solutions exist, are unique, and possess certain regularity properties (cf. Giaquinta (1983), and Giusti (1983)). The uniqueness is an immediate consequence of the version of (6.14), (6.15) holding for $G,H \in \mathbb{L}$. A weaker uniqueness result would hold when $\hat{T}(\cdot,x)$ is restricted to $\mathbb{L}^+$.

At first sight these virtues of (6.14), (6.15) seem irresistible. But the very uniqueness could mean that an elastic rod could never buckle, no matter how thin it is and how large the thrust applied to it is. A main reason for suffering the complexity of nonlinear elasticity is to be able to describe buckling. Another objection to (6.1) and (6.1) is that a strictly convex, continuous stored energy function $\phi(\cdot,x)$ cannot blow up on the boundary $\partial\mathbb{L}^+$ of a nonconvex set $\mathbb{L}^+$. Indeed, if we refined (6.12) by a standard assumption used in the theory of monotone operators, namely by requiring that the left side of (6.14) approach $\infty$ as $G+H \longrightarrow \partial\mathbb{L}^+$ along a line lying in $\mathbb{L}^+$, then this refinement is also incompatible with the nonconvexity of $\mathbb{L}^+$. It can be shown that (6.14) and (6.15) are incompatible with (6.3).

The easiest mathematical way to allow nonuniqueness, while preserving most of the analytical advantages of (6.14), (6.15) would be to let $\hat{T}$ depend on p as well as on $\partial p/\partial x$, but we have precluded this ploy on physical grounds.

A condition weaker than (6.14), (6.15), yet having a well-established mathematical status, is the strict form of the <u>strong ellipticity condition</u>: (6.14) holds

(6.18a) $\forall\, H$ of rank 1 such that $G+H \in \mathbb{L}^+$.

(If $\hat{T}(\cdot,x)$ is differentiable, then a slightly stronger version of this condition is that (6.16) hold

(6.18b) $\forall$ H of rank 1. )

This condition ensures that (1.1), (1.2) is hyperbolic and can therefore admit solutions with a full range of wave-like behavior. For hyperelastic materials, the strong ellipticity condition reduces to the strong Legendre-Hadamard condition of the calculus of variations.

Now every tensor H of rank 1 can be written as a scalar multiple of a tensor product $a \otimes b$ of unit vectors a and b. Since $G:(a \otimes b) \equiv a \cdot (Gb) \longmapsto \det G$ is affine for fixed $G - a \cdot (Gb) a \otimes b$, the set $\{\lambda \in \mathbb{R}: G + \lambda a \otimes b \in \mathbb{L}^+\}$ is either empty, a half line, or the whole line. For this reason restrictions on H in (6.18) are milder than those in (6.15). Moreover, (6.14), (6.18) says that the component of stress in one direction is an increasing function of the increment in the corresponding component of $\partial p/\partial x$. In particular, if we replace H in (6.16) with $a \otimes b$, then we obtain

$$\frac{\partial a \cdot [\hat{T}(G, x) b]}{\partial a \cdot (Gb)} > 0. \tag{6.19}$$

This inequality gives a specific and attractive interpretation to criterion (ii).

In summary, the monotonicity condition is physically unreasonable while the strong ellipticity seems satisfactory. Erickson (1983), however, has persuasively argued that the strong ellipticity condition should not be regarded as universally valid in nonlinear elasticity because there are crystalline materials that violate it. In this setting, existence, regularity and multiplicity theorems can help illuminate the issues involved. We now turn to this question.

## 7. Existence Theories.

Hughes, Kato, and Marsden (1977) proved the existence of solutions of the Cauchy problem for the equations of motion of nonlinear elasticity for short time intervals when the material satisfies the strong ellipticity condition. John (1977, 1983) has studied the propagation of waves and the life-span of solutions for hyperbolic systems including the equations of nonlinear elasticity. The settings of these works are such that the difficulty with (2.2) does not intervene in the analysis.

The situation for static problems is consequently more delicate. The main results are for hyperelastic materials, in which case the governing equations are the Euler-Lagrange equations for the functional

$$p \longmapsto \int_{\Omega} \phi\left[ \frac{\partial p}{\partial x}(x), x \right] dx + \cdots \tag{7.1}$$

where the ellipsis refers to potentials of any external forces applied to the body.

Morrey (1952, 1966) introduced the strong quasiconvexity condition, which implies the strong ellipticity condition, and showed that functionals meeting this nonlocal condition together with certain growth conditions have minimizers on appropriate Sobolev spaces. This remarkable result was incapable of handling (2.2) and (6.13). Ball (1977a,b) was able to overcome this difficulty. He said that $\phi(\cdot,x)$ is <u>polyconvex</u> if there is a convex function $\Theta(\cdot,x):\mathbf{L}^+ \times \mathbf{L}^+ \times (0,\infty) \longrightarrow \mathbb{R}$ such that

$$\phi(G,x) = \Theta(G,G^{\times},\det G,x). \tag{7.2}$$

Here $G^{\times}$ is the cofactor tensor of G. Ball showed that the convex functions form a proper subset of the polyconvex functions, which in turn form a proper subset of the strongly quasiconvex functions. Note that polyconvexity is a local condition. He then showed that the polyconvexity allows multiplicity of solutions. He finally showed that if $\phi$ is polyconvex, then (7.1) is minimized on subsets of Sobolev spaces satisfying (2.2) in an appropriate sense.

Ball was not able to show that the minimizers are weak solutions of the Euler-Lagrange equations. This is not surprising. The usual way to show that minimizers are weak solutions is to impose bounds on $\partial\phi/\partial G$ having the effect that the mapping from p to $x \longmapsto (\partial\phi/\partial G)((\partial p/\partial x)(x),x)$ takes elements from the Sobolev space in which p lies to its dual space. These bounds are used to justify the existence of the Gâteaux derivative by the use of the dominated convergence theorem or one of its relatives. But the growth of $\phi$ at $\partial\mathbb{L}^+$ precludes such bounds. Indeed, proofs that minimizers are weak solutions when (6.13) holds have been attained only for ordinary differential equations in ways that depend crucially on the presence of but a single independent variable. (Cf. Antman (1970, 1976), Antman and Brezis (1978).) Moreover, in many respects it is easier to prove the existence of weak solutions directly by using the theory of operators of monotone type (for ordinary differential equations of elasticity) than to prove that a minimizer is a weak solution. (Cf. Antman (1983).) Thus, it is not inconceivable that there is an obstruction to the proof that minimizers are weak solutions of the partial differential equations.

Ball (1980,1981a,b) has also described the kinds of irregularities possible in materials satisfying the strong ellipticity condition and the polyconvexity condition.

Further discussions of the analytic consequences of constitutive relations are given by Ball (1977) and Antman (1983). These articles have extensive bibliographies.

## 8. A Comparison of Solid Mechanics with Fluid Mechanics.

A poll would show that the number of scientists with a valid impression of the equations of fluid dynamics far exceeds the number with such an impression of the equations of nonlinear elasticity. Here we sketch the manifold distinctions between these two theories to show why this state of affairs obtains.

We suppose that for each t the function $p(\cdot,t)$ is one-to-one. Let $p(\Omega,t) \ni y \longmapsto q(y,t) \in \Omega$ be its inverse; $q(y,t)$ is the material point occupying position y at time t. We set

$$(8.1) \quad v(y,t) = \frac{\partial p}{\partial t}(q(y,t),t), \qquad a(y,t) = \frac{\partial^2 p}{\partial t^2}(q(y,t),t).$$

$v(y,t)$ is the velocity and $a(y,t)$ is the acceleration of the material point occupying position y at time t. By the chain rule (applied to $v(p(x,t),t) = (\partial p/\partial t)(x,t)$ ) we obtain

$$(8.2) \quad a(y,t) = \left[ \frac{\partial v}{\partial y}(y,t) \right] v(y,t) + \frac{\partial}{\partial t} v(y,t).$$

When the independent spatial variable is x, which lies in a fixed domain $\Omega$, a problem in continuum mechanics is said to be given a <u>material</u> (or a <u>Lagrangian</u>) description (which was introduced by Euler). This is the most natural description from the viewpoint of physics. When the independent spatial variable is y, which lies in $p(\Omega,t)$, a problem is said to be given a <u>spatial</u> (or <u>Eulerian</u>) description (which was introduced by D'Alembert). Such descriptions are used primarily for problems of fluid dynamics for which $p(\Omega,t)$ is fixed. The awkwardness of this description, manifested by the nonlinearity in the expression (8.2) for the acceleration, is compensated for by a variety of analytical advantages.

A fluid is incompressible if the volume of any part of it cannot be changed, no matter what system of forces is applied to it; it is accordingly subject to the constraint that

$$(8.3) \qquad \det(\partial p/\partial x) = 1.$$

(Water is effectively imcompressible; gases at low speeds are approximately so.) Differentiating (8.3) with respect to t and using (8.1), we obtain after some computation that

(8.4) $$\operatorname{div} v \equiv \operatorname{trace}(\partial v/\partial y) = 0.$$

That (8.4) is linear, whereas (8.3) is not, is one advantage of the spatial description.

Now we introduce the <u>Cauchy stress tensor</u> $\Sigma$ (y,t) which measures force per unit <u>actual</u> area across a fixed surface in space at the material point occupying position y at time t. A uniform incompressible fluid is <u>Newtonian</u> if it has a constitutive equation of the form

(8.5) $$\Sigma\,(y,t) = -\,\pi(y,t)I + \mu\left[\frac{\partial v}{\partial y}(y,t) + \left[\frac{\partial v}{\partial y}(y,t)\right]^{*}\right]$$

where I is the identity tensor, $\pi$ is a Lagrange multiplier, called the <u>pressure</u>, maintaining the constraint (8.4), and $\mu$ is a positive number called the <u>viscosity</u>.

The equations of motion for $\Sigma$ are

(8.6) $$\operatorname{div} \Sigma + g = \sigma a$$

where g is the given force per unit actual volume and $\sigma$ is the mass density per unit actual volume. (For uniform incompressible materials $\sigma = \rho$.) The substitutions of (8.2) and (8.5) into (8.6) yield the <u>Navier-Stokes equations</u>

(8.7) $$\mu\Delta v - \frac{\partial \pi}{\partial y} + g = \sigma\left[\frac{\partial v}{\partial y}\,v + \frac{\partial v}{\partial t}\right]$$

where $\Delta$ is the Laplacian. Equation (8.7) is to be supplemented with (8.4).

If the fluid were compressible, then (8.4) would be replaced by an equation involving $\sigma$ and v expressing conservation of mass. A constitutive function would give $\pi$ in terms of $\sigma$. This function would reflect the elastic response of the fluid. We limit our attention to incompressible fluids because we wish to contrast elastic response with viscous response.

The Navier-Stokes equations describe a great many common fluids, such as water. Other fluids, called non-Newtonian, exhibit some strange mechanical effects. Paint, sour milk, and an assortment of industrial pollutants are non-Newtonian.

| | | Elastic Solids | Incompressible Viscous Fluids |
|---|---|---|---|
| Type of Equation | Linear | (1) ////////////// | (2) |
| | Semilinear | (3) | (4) ////////////// |
| | Quasilinear | (5) ////////////// | (6) |

Table 8.8. Theories of elasticity and of incompressible, viscous fluids. Theories deemed to be well-established occupy boxes that are hatched.

In table 8.8 we indicate those theories of elasticity and of incompressible, viscous fluids that are well-established, possessing precisely formulated equations of broad applicability. Box 1 corresponds to the linear theory of elasticity, which describes the very small deformations of solids and which has a complete and elegant theory (cf. Gurtin (1972), Fichera (1972).) Linear elasticity is the basic theory for many problems of structural design.

Box 4 corresponds to Newtonian fluids whose theory we have just described. It is analogous to linear elasticity in that the constitutive equations for each theory are linear. The convective nonlinearity in the acceleration, which renders the governing Navier-Stokes equations semilinear, is the source of great technical difficulty. (In sensible formulations of solid mechanics the artifice of a spatial description is avoided, so that difficulties with the convective nonlinearity do not arise. Box 2 corresponds to the Stokes equations, obtained by discarding the convective term from the Navier-Stokes equations. It describes slow flow. It and related linear theories, which are analogous to linear elasticity, have a limited utility relative to that of the Navier-Stokes equations.

Box 3 is empty because the basic nonlinearity of (2.4), (2.5) precludes a choice of constitutive function leading to semilinear equations. There are a variety of semilinear equations for elastic structures, but these for the most part are obtained by ad hoc approximations.

Box 5 corresponds to the exact theory of nonlinear elasticity, which is the main subject of this paper.

There are several candidates for the non-Newtonian fluids of the Box 6. The Reiner-Rivlin fluid is obtained by replacing (8.5) with the prescription of $\Sigma + \pi I$ as a nonlinear isotropic function of the symmetric part of $\partial v/\partial y$. But certain effects predicted by this theory are not observed in any real fluids. The more general Rivlin-Ericksen theories allow $\Sigma + \pi I$ to depend in an invariant way on $\partial v/\partial y$, $\partial a/\partial y$, etc. But many of these theories produce strange stability results. Analogous theories are available for solids (which are called viscoelastic). More generally, there are theories for both solids and fluids in which the partial differential equations are replaced by functional partial differential equations (obtained by prescribing the stress at the present time to be a function of the past history of C).

In summary, fluid dynamics has a single, very useful semilinear system, the Navier-Stokes equations, parametrized by a single scalar, the viscosity. Solid mechanics has a useful quasilinear system, the equations of nonlinear elasticity, parametrized by nonlinear constitutive functions. No quasilinear system for fluid dynamics is as well established on analytical or mechanical grounds. (For an interesting, encyclopedic treatment of the issues hinted at in this section, see Truesdell and Noll (1965). )

**Acknowledgment.** The preparation of this paper was partially supported by NSF Grant MCS-80-01844. The paper was written at the Mathematical Sciences Research Institute, Berkeley, California.

## References

J.C. ALEXANDER and S.S. ANTMAN (1982), The ambiguous twist of Love, Quart. Appl. Math., 40, 83–92.

S.S. ANTMAN (1971), Existence and nonuniqueness of axisymmetric equilibrium states of nonlinearly elastic shells, Arch. Rational Mech. Anal., 40, 329–371.

S.S. ANTMAN (1976), Ordinary differential equations of one-dimensional nonlinear elasticity, Arch. Rational Mech. Anal., 61, 307–393.

S.S. ANTMAN (1983), Regular and singular problems for large elastic deformations of tubes, wedges, and cylinders, Arch. Rational Mech. Anal., 83, 1–52. Corrections to appear.

S.S. ANTMAN and H. BREZIS (1978), The existence of orientation-preserving deformations in nonlinear elasticity, in Nonlinear Analysis and Mechanics, edited by R. Knops, Pitman Res. Notes in Math. 27, London, 1–29.

S.S. ANTMAN and J.E. OSBORN (1979), The principle of virtual work and integral law of motion, Arch. Rational Mech. Anal., 69, 231–262.

J.M. BALL (1977a), Convexity conditions and existence theorems in nonlinear elasticity, Arch. Rational Mech. Anal. 63, 337–403.

J.M. BALL (1977b), Constitutive inequalities and existence theorems in nonlinear elastostatics, in Nonlinear Analysis and Mechanics: Heriot-Watt Symposium, Vol. I, Pitman Research Notes in Math., 17, London, 187–241.

J.M. BALL (1980), Strict convexity, strong ellipticity, and regularity in the calculus of variations, Math. Proc. Camb. Phil. Soc., 87, 501–513.

J.M. BALL (1981a), Global invertibility of Sobolev functions and the interpenetration of matter, Proc. Roy. Soc. Edinburgh 88A, 315–328.

J.M. BALL (1981b), Remarques sur l'existence et la régularité des solutions d'élastostatique nonlinéaire, in Recent Contributions to Nonlinear Partial Differential Equations, edited by H. Berestycki and H. Brezis, Pitman, London.

J.M. BALL, ed. (1983), Systems of Nonlinear Partial Differential Equations, Reidel, Dordrecht.

J.L. ERICKSEN (1983), Ill-posed problems in thermoelasticity theory, in Ball (1983).

G. FICHERA (1972), Existence Theorems in Elasticity, in Handbuch der Physik, Vol. VIa/2, Springer-Verlag, 347-389.

M. GIAQUINTA (1983), The regularity problem of extremals of variational integrals, in Ball (1983).

E. GIUSTI (1983), Some aspects of the regularity theory for nonlinear elliptic systems, in Ball (1983).

M.E. GURTIN (1972), The Linear Theory of Elasticity, in Handbuch der Physik, Vol. VIa/2, Springer-Verlag, 1-295.

M.E. GURTIN and L.C. MARTINS (1976), Cauchy's theorem in classical physics, Arch. Rational Mech. Anal. 60, 305-324.

T.J.R. HUGHES, T. KATO and J.E. MARSDEN (1977), Well-posed quasilinear second-order hyperbolic systems with applications to nonlinear elastodynamics and general relativity, Arch. Rational Mech. Anal. 63, 273-294.

F. JOHN (1977), Finite amplitude waves in a homogeneous isotropic elastic solid, Comm. Pure Appl. Math. 30, 421-446.

F. JOHN (1983), Lower bounds for the life span of solutions of nonlinear wave equations in three dimensions, Comm. Pure Appl. Math. 36, 1-35.

C.B. MORREY, Jr. (1952), Quasi-convexity and the lower semicontinuity of multiple integrals, Pacific J. Math. 2, 25-53.

C.B. MORREY (1966), Multiple Integrals in the Calculus of Variations, Springer-Verlag.

C. TRUESDELL (1977), A First Course in Rational Continuum Mechanics, Vol 1, Academic Press.

C. TRUESDELL and W. NOLL (1965), The Non-linear Field Theories of Mechanics, Handbuch der Physik III/3, Springer-Verlag.

# AN INTRODUCTION TO EULER'S EQUATIONS FOR AN INCOMPRESSIBLE FLUID

Alexandre J. Chorin

Euler's equation for a fluid of constant density can be written in the form

$$\text{(1a)} \qquad \frac{D}{Dt}\underline{u} + \text{grad } p = 0,$$

$$\text{(1b)} \qquad \text{div } \underline{u} = 0,$$

where $\underline{u}(\underline{x},t)$ is the velocity at the point $\underline{x}$ at time t, p is the pressure, and $\frac{D}{Dt}$ is the operator $\frac{\partial}{\partial t} + \underline{u}\cdot\nabla$. $\underline{u}(x,0)$ is assumed given, the number of space dimensions is either $n = 2$ or $n = 3$, and we greatly simplify the problem by assuming that there are no boundaries (i.e., either the flow occupies the whole space or it is periodic in all components of $\underline{x}$). The energy $E = \frac{1}{2}\int |\underline{u}|^2\, dV$ is assumed finite at $t = 0$.

Equation (1b) expresses the conservation of mass. Equation (1a) is Newton's law: force = mass × acceleration. Mass does not appear explicitly because we use units in which the density is 1. $\frac{D}{Dt}\underline{u}$ can be readily seen to be the acceleration of a fluid particle; -grad p is the force acting on that particle. This expression for the force arises when one assumes that one part of the fluid exerts on an adjacent part of the fluid a force normal to their surface of contact and proportional to a scalar function p. If $\underline{u}$ is smooth, E = constant is a consequence of (1a), (1b). If $\underline{u}$ is not smooth, one must assume in addition $\frac{dE}{dt} \leq 0$. For a detailed derivation of these equations, see [2], [3].

Equations (1) describe a variety of important and baffling physical phenomena, in particular fully developed turbulence ([16],

[17]). Their mathematical theory is far from settled; in particular, the problem of global existence of smooth solution when $n = 3$ is open ([5], [6]). It is very likely that any light shed on the mathematical issues will help to clarify the physics, and vice-versa.

The assumption that forces are perpendicular to interface implies that rotation is persistent (brakes do not work without tangential forces). The mathematical form of this fact is the Kelvin circulation theorem: Let C be a closed curve immersed in the fluid; let $\Gamma_c$ be the <u>circulation</u> on C,

$$\Gamma_c = \int_C \underline{u} \cdot d\underline{s}.$$

$\Gamma_c$ is a measure of rotation. The circulation theorem states that $\frac{D\Gamma_c}{Dt} = 0$, i.e., $\Gamma_c$ does not change as C is moved by the fluid and as $\underline{u}$ evolves.

The vector $\underline{\xi} = \text{curl } \underline{u}$ is called the vorticity. An integral line of $\underline{\xi}$ is called a vortex line. A surface tangent to $\underline{\xi}$ at each point is called a vortex sheet. One can see from the circulation theorem that vortex lines and vortex sheets remain vortex lines and vortex sheets as the flow evolves. A cylinder-like vortex sheet is called a vortex tube. Let S be a section of a vortex tube; the quantity

$$\kappa = \int_S \underline{\xi} \cdot d\underline{S}$$

is the circulation of the tube. The circulation is independent of the particular section used (by the divergence theorem), and is a constant in time for the moving tube (by the circulation theorem). If $\underline{\xi} = 0$ at $t = 0$, $\underline{\xi} = 0$ for all t.

Equations (1) can be written in terms of $\underline{\xi}$:

$$\frac{D\underline{\xi}}{Dt} - (\underline{\xi} \cdot \underline{u})\underline{u} = 0, \qquad \underline{\xi} = \text{curl } \underline{u}, \qquad \text{div } \underline{u} = 0.$$

Consider three dimensional flow. Vortex tubes can stretch, and if $\underline{\xi} \not\equiv 0$, the constancy of $\kappa$ implies that $\int |\xi|^2 \, dv$ can increase.

It is known from physical experiment and numerical computation that the following facts hold for a flow satisfying equation (1):

(a) The flow is intermittent: The "active" volume, in which $|\underline{\xi}|$ is large, comprises much less than the total volume available to the flow.

(b) $\int |\underline{\xi}|^2 \, dV$ increases very rapidly in time. An unanswered question is: does it become infinite in a finite time? (For possible numerical answers, see [16], [17], [20].)

Fact (a) is readily explained if different tubes stretch at different rates (in fact, because of conservation of mass, if some tubes stretch, others must shrink). Fact (b) is related to the following fact: consider the Navier-Stokes equations (i.e., the equations of flow in which viscous stresses are added in equation (1a)):

$$(2) \qquad \frac{D\underline{u}}{Dt} + \text{grad } p = \nu \, D\underline{u}, \qquad \text{div } \underline{u} = 0, \qquad \nu = \text{viscosity}.$$

The energy dissipation $\epsilon = \frac{dE}{dt}$ can be shown to be proportional to $\nu Q(\frac{d\underline{u}}{dx})$, where Q is a quadratic functional of the derivatives of $\underline{u}$. Experimentally, it appears that as $\nu \longrightarrow 0$, $\epsilon \not\to 0$. If the likely (but not proved) convergence theorems hold in three space dimensions, one would conclude that the derivatives of $\underline{u}$ in (1), (for example, those which enter $\underline{\xi}$), become infinite in a finite time. A similar situation holds for the equation

$$(3) \qquad u_t + \tfrac{1}{2}(u^2)_x = 0, \qquad u(x,0) \text{ smooth and of compact support.}$$

Consider $E = \int u^2 \, dx$. If u is smooth, E = constant. However, the solutions of (3) cease to be smooth in a finite time and $\frac{dE}{dt} < 0$ (see e.g. [21]).

Vortex stretching provides the only plausible mechanism for singularity formation for equations (1). (It has almost been proved that there is no other.) Suppose a vortex tube stretches so that one of its cross-sections becomes of measure 0. Then $|\underline{\xi}|$ becomes infinite over that cross-sections. However, consider the relation between $\underline{u}$ and $\underline{\xi}$. Let $\underline{u} = \text{curl } \underline{\psi}$, with div $\underline{\psi} = 0$ ($\underline{\psi}$ is a vector potential.) We have: $\Delta \underline{\psi} = - \xi$, $\underline{\psi} = - \Delta^{-1} \xi$, $\underline{u} = - \text{curl } \Delta^{-1} \underline{\psi}$. An explicit form of these equations can be obtained (it is called the Biot-Savart law) and it shows that if the cross-section of a tube has become a point the energy E is infinite, which is impossible (see [1]). Other arguments show that the cross-section of a tube cannot flatten out to a surface-like object; the only plausible conclusion is that if $\int \underline{\xi}^2$ becomes infinite (in a finite or infinite time), the cross-section of a tube must converge to a fractal object (i.e., an object whose Hausdorff dimension is not an integer). This fractal object presumably arises as a result of a multiple folding of the tube, which can be exhibited on the computer, and which appears to be the only way one can reconcile the energy law $\frac{dE}{dt} \leqslant 0$ with a growth of $\int \underline{\xi}^2$. For a discussion of these facts, and an analogy with potential theory, see [16]. The exceptional set on which the vorticity is very large is an "active region". Different arguments which show that these "active regions" are fractal sets have been given in [19], [18]. Fractal sets also appear in the theory of the Navier-Stokes equations (see c.f. [14], [15]). An interesting question is: What is the geometrical nature of an exceptional set which can lead to $\frac{dE}{dt} < 0$? Or, in other words, what is the incompressible analogue of the entropy condition which singles out the one correct solution of (1)? (If indeed there is one.)

Note that the discussion just given relies mostly on experimental and numerical results, and that a precise mathematical theory is unavailable. In dimension 2, vortex stretching cannot occur, since all vortex tubes are perpendicular to the plane of the motion. The flow is known to be smooth for all time if the data are smooth. In the absence of boundaries, the limit $\nu \longrightarrow 0$ in (2) is well behaved.

The limit $\nu \longrightarrow 0$ in equations (2) and its relation to the solution of (1) in the presence of boundaries present a variety of hard and unsolved problems, of great physical significance, in particular for aerodynamics and the theory of flight (see e.g. [9]).

## Selected References

General Introductions to Euler and Navier-Stokes Equations:

[1] G. K. Batchelor, An introduction to fluid mechanics, Cambridge (1967).

[2] A. Chorin, J.E. Marsden, A mathematical introduction to fluid mechanics, Springer (1979).

[3] L. Landau and E. Lifshitz, Fluid mechanics, Pergamon (1969).

[4] R. Temam, Navier-Stokes equation and nonlinear functional analysis, lecture notes, Orsay (1982).

Existence for Euler's Equations

[5] T. Kato, Arch. Rat. Mech. Anal. 25, 188 (1967).

[6] R. Temam, J. Funct. Anal. 20, 32 (1975).

The Limit $\nu \longrightarrow 0$

[7] T. Beale and A. Majda, Math Comp. 37, 243 (1982).

[8] D. Ebin and J. Marsden, Ann. of Math 92, 102 (1970).

[9] S. Goldstein, Modern developments in fluid dynamics, Dover (1938, 1965), Chapters 1 and 2.

Introduction to Turbulence Theory

[9] above and,

[10] G. K. Batchelor, The theory of homogeneous turbulence, Cambridge (1961).

[11] P. Bernard and T. Ratiu (eds.), Berkeley turbulence seminar, Springer (1978).

[12] A. Chorin, Lectures on turbulence theory, Publish/Perish, Boston (1975).

[13] R. Temam, Turbulence and NS Equation, Springer (1975).

Existence Theory for the Navier Stokes Equations Involving Exceptional Sets

[14] V. Schaffer, Comm. Math. Phys. 55, 97 (1977).

[15] L. Caffarelli, R. Kohn, L. Nirenberg, Comm. Pure Appl. Math. 25, 771 (1981).

Vortex Stretching, Exceptional Sets and the Blow-up Problem

[16] A. Chorin, Comm. Pure Appl. Math 34, 853 (1981).

[17] A. Chorin, Comm. Math. Phys. 83, 517 (1982).

[18] U. Frisch, P. Salem, M. Nelkin, J. Fluid Mech. 87, 719 (1978).

[19] B. Mandelbrot, J. Fluid Mech. 62, 331 (1974).

[20] R. Morf, S. Orszag, U. Frisch, Phys. Rev. Letters 44, 572 (1980).

An Introduction to Hyperbolic Equations

[21] P. Lax, Hyperbolic system of conservation laws and the mathematical theory of shock waves, SIAM Publications (1975).

## Linearizing Flows and a Cohomology Interpretation of Lax Equations.

A talk given by Phillip Griffiths in Chern's Differential Equations Seminar on May 9, 1983

§1. By a Lax equation with a parameter we shall mean an equation

$$\dot{A}(\xi) = [B(\xi), A(\xi)] \tag{1}$$

where

$$\begin{cases} A(\xi) = \sum_{-p}^{q} A_k(t)\xi^k \\ B(\xi) = \sum_{-p}^{q} B_k(t)\xi^k \end{cases}$$

are finite Laurent series in a variable $\xi$ whose coefficients are matrices depending on a parameter. It is known ([2], [3]) that, with the exception of Kowaleski's top, all of the known completely integrable Hamiltonian systems may be represented in the form (1). We give three cases.

**Example** 1. The Euler equations of a free rigid body in $\mathbb{R}^n$ are ([14], [21])

$$\begin{cases} \dot{M} = [M, \Omega] & \text{where} \\ \Omega \in so(n), & M = \Omega J + J\Omega \in so(n) \\ J = \mathrm{diag}(\lambda_1, \ldots, \lambda_n), & \lambda_i > 0 \end{cases} \tag{2}$$

By Manakov's trick these are equivalent to (1) where

$$\begin{cases} A = M + J^2\xi \\ B = -(\Omega + J\xi) \end{cases}$$

**Example 2.** The Toda lattice is the Hamiltonian system on $T^*\mathbb{R}^n \cong \mathbb{R}^{2n}$ corresponding to Hamiltonian function $H(x,y) = \frac{1}{2}\Sigma y_i^2 + \sum_{i=1}^{n} e^{x_i - x_{i+1}}$, $x_{n+1} = x_1$. By Flaschka's substitution

$$\begin{cases} a_k = \frac{1}{2} e^{(x_k - x_{k+1})} \\ b_k = -y_k/2 \end{cases}$$

the Hamiltonian equations are

$$(3) \qquad \begin{cases} \dot{b}_k = 2(a_k^2 - a_{k+1}^2), & a_{n+1} = a_1 \\ \dot{a}_k = a_k(b_{k+1} - b_k), & b_{n+1} = b_1. \end{cases}$$

Noting that $\sum_k b_k$ = constant, we normalize by requiring that $\sum_k b_k = 0$. Then by [24] the equations (3) are of the form (1) where

$$A(\xi) = \begin{bmatrix} b_1 & a_1 & a_n\xi^{-1} \\ a_1 & & \\ & & a_{n-1} \\ a_n & a_{n-1} & b_n \end{bmatrix}$$

$$B(\xi) = \begin{bmatrix} 0 & a_1 & -a_n\xi \\ -a_1 & & \\ & & a_{n-1} \\ a_n\xi & -a_{n-1} & 0 \end{bmatrix}$$

**Example** 3. Nahm's equations [20], which arise in the study of monopoles, are

$$\begin{cases} \dot{T}_i = \frac{1}{2} \epsilon_{ijk} [T_j, T_k] \\ T_i \in u(n). \end{cases}$$

By [10] these are equivalent to (1) where

$$\begin{cases} A(\xi) = (T_1 + iT_2) + (-2iT_3)\xi + (T_1 - iT_2)\xi^2 \\ B(\xi) = -A_\xi = -\frac{A_1}{2} - \xi A_2. \end{cases}$$

§2. Given (1) we define its spectral curve C to be the normalization of the complete algebraic curve whose affine equation is

$$Q(\xi,\eta) = \det\|\eta I - A(\xi,t)\| = 0.$$

Since the Lax equations give isospectral flows, C is independent of t. We assume that for general $p = (\xi,\eta)$ the corresponding eigenspace is 1-dimensional and is spanned by a vector $v(p,t) \in V \cong \mathbb{C}^n$. There is then a family of holomorphic mappings

$$f_t\colon C \longrightarrow \mathbb{P}V \tag{4}$$

given by $p \longmapsto \mathbb{C}v(p,t)$ ; clearly these give the time evolution of $A(\xi,t)$. We set

$$L_t = f_t^*(\mathcal{O}_{\mathbb{P}V}(1)) \in J(C)$$

where $\mathcal{O}_{\mathbb{P}V}(1)$ is the hyperplane line bundle and

$$J(C) = H^1(\mathcal{O}_C)/H^1(C,\mathbb{Z}) \tag{5}$$

$$\cong H^0(\Omega_C)^*/H_1(C,\mathbb{Z}) \tag{6}$$

is a complex torus giving the Jacobian variety of C (c.f. [4] for definitions). Thus associated to (1) is the flow

$$t \longrightarrow L_t \in J(C) \tag{7}$$

and motivated by [2], [3] we may consider the following

(8) **<u>Problem</u>**: <u>Determine the necessary and sufficient conditions on</u> $B(\xi)$ <u>that the flow (7) be linear</u>.

Now, for an arbitrary family of homomorphic mappings (4), reasonably standard deformation theory ([4]) may be used to answer this problem. Moreover, since the tangent space to any algebro-geometric moduli space is computed cohomologically, the general answer to (8) is expressed in terms of an $H^1$ (by (5) this is reasonable).

Suppose now that (4) arises from (1). Note that B is not unique since any substitution

$$B \longmapsto B + P(\xi, A), \qquad P(\xi,\psi) \in \mathbb{C}[\xi,\psi],$$

leaves (1) invariant. This suggests that the B in a Lax pair (A,B) lives naturally in a cohomology group somewhere. By a very nice cohomological computation, this turns out to be the case and allows us to answer (8) in a way that is effective for the computation of examples.

To explain the result we assume for simplicity that

$$B(\xi) = \sum_{k=0}^{\ell} B_k(t)\xi^k$$

is a polynomial of degree $\ell$. View $\xi$ as a meromorphic function, set

$$D = \xi^{-1}(\infty) = \sum_i n_i p_i$$

and denote by $H^0(\mathcal{O}_D(P))$ the Laurent tails $\{\phi_i\}$ where

$$\phi_i = a_{i,n_i}/z_i + \cdots + a_{i,1}/z_i$$

and where $z_i$ is a local coordinate around $p_i$. Near $p_i$ we have

$$A\, v(p,t) = \eta\, v(p,t)$$

$$\Rightarrow \quad \dot{A}v + A\dot{v} = \eta\dot{v}$$

$$\Rightarrow \quad A(\dot{v} - Bv) = \eta(\dot{v} - Bv) \qquad \text{(by (1))}$$

(9) $$\Rightarrow \quad Bv = \dot{v} + \lambda_i v,$$

where $\lambda_i$ is a Laurent tail as above.

**Definition**. We defined the residue $\rho(B) \in H^0(\mathcal{O}_D(D))$ to be the collection of Laurent tails $\{\lambda_i\}$ given by (9).

(10) **Theorem**. *Let $\mathcal{L} \subset H^0(\mathcal{O}_D(D))$ be the Laurent tails of meromorphic functions $g \in H^0(\mathcal{O}_C(D))$. Then the flow (7) is linear $\Longleftrightarrow$*

$$\dot{\rho}(B) \equiv 0 \bmod \{\rho(B), \mathcal{L}\}. \tag{11}$$

*If this is satisfied, then using (6) it is given by a translate of*

$$(t,\omega) \longmapsto t \sum \operatorname{Res}_{p_i}(\lambda_i \omega), \qquad \omega \in H^0(\Omega_C). \tag{12}$$

§3. This result serves to unify the known linearizibility theorems given in [2], [3], [10], [17], [21] and [22]. We shall indicate how it applies in two of the above examples.

**Example 1 reconsidered.** In this case

$$D = \xi^{-1}(\infty) = \sum_i p_i$$

and setting $z_i = \xi^{-1}$ near $p_i$

$$\rho(B) = \sum_i \lambda_i/z_i$$

Thus $\dot{\rho(B)} = 0$ and so (11) and (12) apply.

Actually, in this case since $A = M + J^2\xi$ where ${}^tM = -M$ and ${}^tJ^2 = J^2$ we have that

$$Q(\xi,\eta) = (-1)^n Q(-\xi,-\eta).$$

Thus $j(\xi,\eta) = (-\xi,-\eta)$ gives an involution of C with quotient curve C' = C/j. If C and C' have respective genera g and g', then an easy computation using the Riemann-Hurwitz formula gives

$$g - g' = \frac{1}{2}\left(\frac{n(n-1)}{2} - [\tfrac{n}{2}]\right). \tag{13}$$

Since clearly

$$j(\rho(B)) = -\rho(B)$$

the flow (7) actually occurs on the complex torus

$$\mathrm{Prym}(C/C') =: H^0(\Omega_C)^- / H_1(C,\mathbb{Z})^-$$

where $^-$ denotes the $-1$ eigenspace of j. By (13)

$$\dim \mathrm{Prym}(C/C') = \frac{1}{2}\left(\frac{n(n-1)}{2} - [\tfrac{n}{2}]\right). \tag{14}$$

On the other hand, in the Euler equation (2), $\Omega$ moves on an adjoint orbit $\mathcal{O}_{\mu} \subset so(n)$ and in general

(15) $$\dim \mathcal{O}_{\mu} = \frac{n(n-1)}{2} - [\tfrac{n}{2}].$$

Comparing (14) and (15) we see that our linearization occurs on a torus of exactly the correct dimension.

**Example 3 reconsidered**. In this case also

$$D = \sum_i p_i.$$

Near $p_i$ we have

$$\begin{cases} (A_2\xi^2)v_i = \eta_i v_i + O(\xi) \\ Bv_i = -A_2\xi v_i + O(1) \end{cases}$$

$$\Rightarrow \rho^{(B)} = \sum_i \lambda_i/z_i$$

$$\text{where } \lambda_i = (\eta^2/\xi)(p_i).$$

Clearly then (11) is satisfied and (12) linearizes Nahm's equations (cf. [10] for an extensive discussion).

The remaining integrable systems, such as Toda lattice, heavy symmetric top ([22],[23]), geodesics on an ellipsoid ([17],[12]), and Neumann's mechanical problem ([17],[13]) may be treated in a similar way. The details may be found in an upcoming paper by the author.

[1] R. Abraham and J. Marsden, Foundations of mechanics, Benjamin/Cummings (1978).

[2] M. Adler and P. van Moerbeke, Completely integrable systems, Euclidean Lie algebras, and curves, Advances in Math., 38 (1980), 267-317.

[3] M. Adler and P. van Moerbeke, Linearization of Hamiltonian systems, Jacobi varieties, and representation theory, Advances in Math., 38 (1980), 318-379.

[4] E. Arbarello, M. Cornalba, P. Griffiths and J. Harris, Topics in the geometry of algebraic curves, Grundlehren Math., Springer-Verlag, (1984).

[5] V.I. Arnold, Mathematical methods of classical mechanics, Springer-Verlag (1978).

[6] B.A. Dubrovnin, V.B. Mateev, and S.P. Novikov, Uspehi Mat. Nauk., 31 (1976), and Russian Math Surveys, 1 (1976).

[7] B.A. Dubrovnin, Theta functions and non-linear equations, Usp. Mat. Nauk., 36 (1981), 11-80.

[8] I. Frankel, A. Reiman, and M. Senenov-Tian-Shansky, Graded Lie algebras and completely integrable dynamical systems, Soviet Math. Dokl., 20 (1979), 811-814.

[9] R. Hartshorne, Algebraic geometry, Springer-Verlag.

[10] N. Hitchin, On the construction of monopoles, preprint (1982).

[11] E. Horikawa, On deformations of quintic surfaces, Invent. Math., 31 (1975), 43-85.

[12] H. Knörrer, Geodesics on the ellipsoid, Invent. Math., 59 (1980), 119-144.

[13] H. Knörrer, Geodesics on the ellipsoid, Invent. Math., 59 (1980), 119-144.

[14] S.V. Manakov, Note on the integration of Euler's equations of the dynamics of an n-dimensional rigid body, Functional Anal. and

App., 11 (1976), 328-329.

[15] A.S. Mishchenko and A.T. Fomenko, Euler equations on finite dimensional Lie groups, Math. USSR Izvestija, 12 (1978), 371-389.

[16] H.P. McKean, Integrable systems and algebraic curves, Global Analysis, Lecture Notes in Math. number 755 (1979), 83-200.

[17] J. Moser, Various aspects of integrable Hamiltonian systems, CIME, Bressanone, Progress in Math., 8 (1980), Birkhäuser-Boston, 233-289.

[18] D. Mumford, An algebraic-geometrical construction of commuting operators and of solutions to the Toda lattice equation, Proc. Kyoto Conference in algebraic geometry, Publ. Math. Society, Japan (1977).

[19] P. van Moerbeke and D. Mumford, The spectrum of difference operators and algebraic curves, Acta Math., 143 (1979), 93-154.

[20] W. Nahn, All self-dual multimonopoles for all gauge groups, preprint CERN (1981).

[21] T. Ratiu, The motion of the free n-dimensional rigid body, Indiana Univ. Math. J., 24 (1980), 609-629.

[22] T. Ratiu, Euler-Poisson equations on Lie algebras and the N-dimensional heavy rigid body, Amer. J. Math., 104 (1982), 409-448.

[23] A. Reiman and M. Semernov-Tian-Shansky, Reduction of Hamiltonian systems, affine Lie algebras, and Lax equations, Invent. Math., 54 (1979), 81-101.

[24] P. van Moerbeke, The spectrum of Jacobi matrices, Invent. Math., 37 (1976), 45-81.

[25] J.L. Verdier, Algèbres de Lie, systèmes Hamiltoniens, courbes algèbriques, Sem. Bourbaki 566 (1980/81).

## THE RICCI CURVATURE EQUATION

by Richard Hamilton

### 1. The Equation

Let $g = \{g_{ij}\}$ be a Riemannian metric on a manifold M of dimension n. Its Ricci curvature $Rc(g) = \{R_{ij}\}$ is given by the formula

$$R_{ij} = \frac{1}{2(n-1)} g^{k\ell} \left[ \frac{\partial^2}{\partial x^i \partial x^k} g_{j\ell} + \frac{\partial^2}{\partial x^j \partial x^\ell} g_{ik} - \frac{\partial^2}{\partial x^i \partial x^j} g_{k\ell} - \frac{\partial^2}{\partial x^k \partial x^\ell} g_{ij} \right] + \frac{1}{(n-1)} g^{k\ell} g_{pq} \left[ \Gamma^p_{ik} \Gamma^q_{j\ell} - \Gamma^p_{ij} \Gamma^q_{k\ell} \right]$$

where $\Gamma^\ell_{ij}$ is the Christoffel symbol

$$\Gamma^\ell_{ij} = \frac{1}{2} g^{k\ell} \left[ \frac{\partial}{\partial x^i} g_{jk} + \frac{\partial}{\partial x^j} g_{ik} - \frac{\partial}{\partial x^k} g_{ij} \right].$$

We have included the factor (n-1) so that $R_{ij} = g_{ij}$ on the unit sphere $S^n$. We shall consider the equation $Rc(g) = h$ where $h = \{h_{ij}\}$ is a given symmetric tensor on M and we wish to solve for the unknown metric g. This is a second order system of quasilinear partial differential equations. The non-linearities arise from the inverse $g^{ij}$ of the matrix $g_{ij}$ and the terms quadratic in the first derivatives. Since there are the same number of equations as unknowns the system is determined. If we multiply the metric g by a constant c the Ricci tensor is unchanged, so that

$$Rc(cg) = Rc(g).$$

Hence uniqueness of the solution of $Rc(g) = h$ will hold only up to a constant multiple. Likewise existence will require an extra constant to replace the one lost to the homogeneity, so that for a given h we will try to find a constant $c > 0$ and a metric g with $Rc(g) = ch$.

## 2. Dimension Two

To warm up we consider the equation $Rc(g) = h$ on the sphere $S^2$ (see DeTurck [1]). In dimension two we always have $Rc(g) = Rg$ where $R = \frac{1}{n} g^{ij} R_{ij}$ is the scalar curvature. We include the factor n so $R = 1$ on the unit sphere. This shows that if $Rc(g) = h$ then g must be conformal to h. To simplify matters we assume h is a positive symmetric tensor, which we write as $h > 0$. We can then regard h as a metric also. Let $\mu = (\det g)^{1/2}$ and $\nu = (\det h)^{1/2}$ be the volume elements, and

$$V(g) = \int \mu \qquad\qquad V(h) = \int \nu$$

be the total volumes. If $Rc(g) = h$ then $Rg = h$ and $R\mu = \nu$. Hence by the Gauss-Bonnet theorem on $S^2$

$$V(h) = \int \nu = \int R\mu = 4\pi.$$

This imposes a necessary condition on h to be a Ricci tensor. It is also sufficient.

**2.1 THEOREM.** If $h > 0$ on $S^2$ and $V(h) = 4\pi$ then there exists a unique metric g on $S^2$ with $V(g) = 4\pi$ and $Rc(g) = h$.

__Proof__. Since g is conformal to h we look for a solution $g = e^{2f}h$ for some function f. If S is the scalar curvature of h and $\Delta$ is the Laplace operator in h then

$$R = e^{-2f}(S - \Delta f).$$

In order to make $Rc(g) = Rg = h$ we need to solve $S - \Delta f = 1$. But we can solve $\Delta f = S - 1$ if and only if $S - 1$ has mean value zero, which happens if and only if

$$V(h) = \int 1\nu = \int S\nu = 4\pi$$

using Gauss-Bonnet again on h this time.

Note that the theorem says that given any metric h on $S^2$ we can find a positive constant c such that Rc(g) = ch for some metric g on $S^2$, and g is unique up to a constant multiple. It follows as a corollary that any surface whose universal cover is $S^2$ also admits a metric of positive Ricci curvature. (Of course the only example is the projective plane $P^2$.) The proof is that any metric h on the surface lifts to $S^2$. By multiplying h by a constant, we can write Rc(g) = h for some g on $S^2$. The solution g is unique up to a constant, and so is determined by its volume. Therefore g is invariant under the group of covering transformations, so g is a metric on the quotient surface with Rc(g) = h.

We would like to prove the corresponding theorem on $S^3$. If it were true, we could then show in the same way that any three-manifold covered by the sphere had a metric of positive Ricci curvature. Then by the result of [3] we could deform it to a metric of constant positive sectional curvature. This would prove the spherical space form conjecture. Unfortunately, as we shall see later, the prospects for the solution of Rc(g) = h on $S^3$ are more limited that on $S^2$. Nevertheless this is what motivated our original interest in the problem, and it shows that a sufficiently good understanding of the solution of the Ricci curvature equation on $S^3$ could produce important topological results.

### 3. A Variational Principle

The Ricci curvature equation can be cast in variational form. Given a symmetric tensor $h = \{h_{ij}\}$ we define a function $J_h$ on the space of metrics $g = \{g_{ij}\}$ by

$$J_h(g) = \int \mathrm{Tr}_g \left[Rc(g) - h\right] \mu = \int g^{ij} \left[R_{ij} - h_{ij}\right] \mu$$

where $\mathrm{Tr}_g$ denotes the trace with respect to g and $\mu$ is the volume form of g.

**3.1 THEOREM.** We have $Rc(g) = h$ if and only if $g$ is a critical point of $J_h$.

<u>Proof</u>. Let $\tilde{g}$ be an infinitesimal variation in $g$. Then the variation $\tilde{R}_{ij}$ in the Ricci tensor is given by

$$\tilde{R}_{ij} = \partial_p \tilde{\Gamma}^p_{ij} - \partial_i \tilde{\Gamma}^p_{pj}$$

where $\tilde{\Gamma}^p_{ij}$ is the variation in the Christoffel symbol. Hence $\tilde{R}_{ij}$ is composed of covariant derivatives and

$$\int g^{ij} \tilde{R}_{ij} \mu = 0.$$

It is then an easy matter to compute the derivative

$$D J_h(g) \tilde{g} = \langle Rc(g) - h, A(g)\tilde{g} \rangle$$

where $A(g)$ is the transformation

$$[ A(g)\tilde{g} ]_{ij} = \frac{1}{2} g^{k\ell} \tilde{g}_{k\ell} g_{ij} - \tilde{g}_{ij}.$$

Since $A(g)$ is invertible, we have $DJ_h(g) = 0$ if and only if $Rc(g) = h$.

**3.2 COROLLARY.** If $Rc(g) = h$ and $h > 0$, then the identity map from $M$ with metric $g$ to $M$ with metric $h$ is harmonic.

<u>First</u> <u>Proof</u>. We can write $J_h(g)$ as

$$J_h(g) = n \int R \mu - \int g^{ij} h_{ij} \mu.$$

The first term is invariant under a diffeomorphism. The second term is the harmonic map energy of the identity map from $(M,g)$ to $(M,h)$. If $J_h(g)$ is critical for all variations of $g$, it is surely so for those that arise from varying $g$ by a diffeomorphism. Hence the harmonic map energy is critical for all variations of the map, so the identity map is harmonic.

**Second Proof.** One can also compute this directly. If $g_{ij}$ and $h_{ij}$ are two metrics on M with Levi-Civita connections $\Gamma^k_{ij}$ and $\Delta^k_{ij}$, their difference $F^k_{ij} = \Delta^k_{ij} - \Gamma^k_{ij}$ is a tensor, and the identity map from M with metric g to M with metric h is harmonic if and only if $g^{ij} F^k_{ij} = 0$. In general,

$$F^\ell_{ij} = \frac{1}{2} h^{k\ell}(\partial_i h_{jk} + \partial_j h_{ik} - \partial_k h_{ij})$$

where $\partial_i$ is the covariant derivative with respect to the connection $\Gamma^k_{ij}$ of $g_{ij}$. Thus the condition $g^{ij} F^\ell_{ij} = 0$ becomes

$$g^{ij}(\partial_i R_{jk} + \partial_j R_{ik} - \partial_k R_{ij}) = 0$$

when $Rc(g) = h$, and this is just the contracted form of the second Bianchi identity.

## 4. Uniqueness For Constant Curvature

It is difficult to obtain uniqueness for the solution of the Ricci curvature equation except in special cases. We prove the following result.

**4.1 THEOREM.** Let h be a metric of constant curvature +1 on the sphere $S^n$. If g is another metric with $Rc(g) = h$ then $g = ch$ for some constant c.

The proof follows from the following interesting identity. Let $g_{ij}$ and $h_{ij}$ be two metrics on a manifold M, let $\Gamma^k_{ij}$ and $\Delta^k_{ij}$ be their Levi-Civita connections and $F^k_{ij} = \Delta^k_{ij} - \Gamma^k_{ij}$ the difference tensor, let $R_{ij}$ be the Ricci curvature tensor of $g_{ij}$ and let $S_{ijk\ell}$ be the Riemannian curvature tensor of $h_{ij}$. Let $\Delta$ be the Laplace operator with respect to $g_{ij}$.

**4.2 LEMMA.** If the identity map from M with metric $g_{ij}$ to M with metric $h_{ij}$ is harmonic and if $E = g^{ij}h_{ij}$ is the harmonic mapping energy density then we have

$$\Delta E = 2\, g^{ik}g^{j\ell}h_{pq}\, F^p_{i\,j}F^q_{k\ell} + 2(n-1)g^{ik}g^{j\ell}R_{ij}h_{k\ell} - 2g^{ik}g^{j\ell}S_{ijk\ell}.$$

**Proof**. We start with a formula for the difference of the Riemannian curvature tensors $R^p_{i\,j\ell}$ of $\Gamma^p_{i\,j}$ and $S^p_{i\,j\ell}$ of $\Delta^p_{i\,j}$. We have

$$S^p_{i\,j\ell} - R^p_{i\,j\ell} = \partial_i F^p_{j\ell} - \partial_j F^p_{i\ell} + F^p_{iq}\, F^q_{j\ell} - F^p_{jq}\, F^q_{i\ell}$$

where $\partial_i$ is covariant differentiation with respect to $F^k_{i\,j}$. We then contract to get $g^{ik}\; g^{j\ell}h_{pk}\left[S^p_{i\,j\ell} - R^p_{i\,j\ell}\right] = g^{ik}g^{j\ell}S_{ijk\ell} - (n-1)g^{ik}g^{j\ell}R_{ij}\; h_{k\ell}$. Now the identity map of (M,g) to (M,h) is harmonic if and only if $g^{ij}F^p_{i\,j} = 0$. Hence when we contract the right-hand side of the formula above for the difference of the curvature tensors with $g^{ik}g^{j\ell}h_{pk}$ the first and third terms vanish. The second yields

$$- g^{ik}g^{j\ell}h_{pk}\; \partial_j F^p_{i\ell} = - \tfrac{1}{2}\, \Delta\, E + g^{ik}g^{j\ell}\; \partial_j h_{kq}\; F^q_{i\ell}$$

because

$$F^\ell_{i\,j} = \tfrac{1}{2}\, h^{k\ell}(\partial_i h_{jk} + \partial_j h_{ik} - \partial_k h_{ij})$$

as can be easily seen using the formula for $\Delta^\ell_{i\,j}$ in a coordinate chart at a point where $\Gamma^\ell_{i\,j} = 0$. The remaining terms simplify using the identity

$$\partial_j h_{kq} = h_{pk}\; F^p_{jq} + h_{pq}\; F^p_{jk}$$

and yield the desired formula.

**Proof of Theorem 4.1**. If $h_{ij}$ has constant curvature +1 then

$$S_{ijk\ell} = h_{ik}h_{j\ell} - h_{i\ell}h_{jk}$$

and if in addition $R_{ij} = h_{ij}$ then

$$(n-1)g^{ik}g^{j\ell} R_{ij}h_{k\ell} - g^{ik}g^{j\ell}S_{ijk\ell} = ng^{ik}g^{j\ell}h_{ij}h_{k\ell} - g^{ik}h_{ik}g^{j\ell}h_{j\ell}.$$

If we choose a basis in which $g_{ij}$ is the identity matrix and $h_{ij}$ is diagonal with eigenvalues $\lambda_i$, we see that

$$ng^{ik}g^{j\ell}h_{ij}h_{k\ell} - g^{ik}h_{ik}g^{j\ell}h_{j\ell} = n \sum_{i=1}^{n} \lambda_i^2 - \left(\sum_{i=1}^{n} \lambda_i\right)^2$$

and

$$n \sum \lambda_i^2 - \left(\sum \lambda_i\right)^2 = \sum_{i<j} \left(\lambda_i - \lambda_j\right)^2 \geqslant 0,$$

with equality holding only when $h_{ij}$ is a multiple of $g_{ij}$. Moreover

$$g^{ik}g^{j\ell}h_{pq} F^p_{ij}F^q_{k\ell} \geqslant 0$$

with equality holding only when $F^p_{ij} = 0$. Then we have $\Delta E \geqslant 0$ on the compact manifold $S^n$, and this forces $E = g^{ij}h_{ij}$ to be a constant function. Thus $\Delta E = 0$, so $h_{ij} = f \cdot g_{ij}$ for some function $f$. Since $E = g^{ij}h_{ij} = nf$ is constant, $f = c$ for some constant $c$, and $h = cg$. This proves the theorem.

We also get a non-existence result

**4.3 THEOREM.** Let $h > 0$ be a metric on a compact manifold $M$ whose sectional curvature is everywhere $< +1$. Then there does not exist any metric $g > 0$ on $M$ with $Rc(g) = h$.

Proof. We apply the argument above to see that $\Delta E > 0$ everywhere, which is impossible since $\int \Delta E = 0$.

If the constant $c$ is very large then the sectional curvature of $ch$ becomes very small. Hence for every $h > 0$ we can find a constant $C$ such that if $c \geqslant C$ then $ch \neq Rc(g)$ for any $g > 0$. It is an interesting open problem to try to bound the Ricci curvature tensors on the other side. Thus given $h > 0$ on $S^n$, can we find an $\epsilon > 0$ such that if $0 < c < \epsilon$ then $ch \neq Rc(g)$ for any $g > 0$ ? This is true on $S^2$, where for every h there is a unique c with $ch = Rc(g)$.

## 5. Existence Near Constant Curvature

We now give a very nice existence result due to DeTurck [1].

**5.1 THEOREM.** Let $\bar{g}$ be a metric on $S^n$ of constant curvature +1, so $Rc(\bar{g}) = \bar{g}$. Then the image of a neighborhood of $\bar{g}$ under the map Rc is a submanifold of codimension one in a neighborhood of $\bar{g}$. For every h near $\bar{g}$ there exists a unique constant c such that $Rc(g) = ch$ for some g near $\bar{g}$, and g is also the unique solution in the neighborhood of $\bar{g}$ if we normalize the volume so $V(g) = V(\bar{g})$.

Proof. This follows from an application of the inverse function theorem. We shall use the Nash-Moser theorem (see [2]), although DeTurck has shown how to avoid it, because the discussion will bring out more of the geometry involved. We examine the linearization of the operator $Rc(g)$, its derivative

$$D\,Rc(g)\,\tilde{g} = \lim_{t\to 0} \left[Rc(g + t\tilde{g}) - Rc(g)\right] / t,$$

which is a second order linear partial differential operator given by $D\,Rc(g)\tilde{g} = \tilde{h}$ where

$$\tilde{h}_{ij} = \frac{1}{2(n-1)}\, g^{k\ell}\left[\partial_i\partial_k\tilde{g}_{j\ell} + \partial_j\partial_\ell\tilde{g}_{ik} - \partial_i\partial_j\tilde{g}_{k\ell} - \partial_k\partial_\ell\tilde{g}_{ij}\right]$$

This is best computed in three steps. A variation $\tilde{g}_{ij}$ in the metric $g_{ij}$ produces a variation $\tilde{\Gamma}^m_{jk}$ in the Levi-Civita connection $\Gamma^m_{jk}$ given by

$$\tilde{\Gamma}^m_{jk} = \frac{1}{2}\, g^{\ell m}\left[\partial_j\tilde{g}_{k\ell} + \partial_k\tilde{g}_{j\ell} - \partial_\ell\tilde{g}_{jk}\right].$$

A variation $\tilde{\Gamma}^m_{jk}$ in a connection $\Gamma^m_{jk}$ produces a variation $\tilde{R}^m_{ijk}$ in its curvature $R^m_{ijk}$ given by

$$\tilde{R}^m_{ijk} = \partial_i \tilde{\Gamma}^m_{jk} - \partial_j \tilde{\Gamma}^m_{ik}.$$

The Ricci curvature $R_{jk} = R^i_{ijk}$ is obtained by contraction, and its variation is $\tilde{R}_{jk} = \tilde{R}^i_{ijk}$. This gives the desired formula for $\tilde{h}_{ij} = \tilde{R}_{ij}$.

We let $\mathcal{V}$ be the space of smooth vector fields on $S^n$ and $\mathcal{M}$ the space of smooth symmetric tensors on $S^n$; the Riemannian metrics on $S^n$ form an open cone $\mathcal{M}^+ \subset \mathcal{M}$. The map Rc is defined

$$\mathrm{Rc}\colon \mathcal{M}^+ \subset \mathcal{M} \longrightarrow \mathcal{M}$$

and its derivative $D\,\mathrm{Rc}(g)\colon \mathcal{M} \longrightarrow \mathcal{M}$ is a linear map of $\mathcal{M}$ to itself. We define some other linear maps. First we define $L(g)\colon \mathcal{V} \longrightarrow \mathcal{M}$ by $L(g)\tilde{v} = \tilde{g}$ where

$$\tilde{g}_{ij} = g_{ik}\partial_j \hat{v}^k + g_{jk}\partial_i \hat{v}^k.$$

This is the classical Lie derivative. An infinitesimal vector field $\hat{v}^k$ produces an infinitesimal diffeomorphism which produces an infinitesimal variation $\tilde{g} = L(g)\tilde{v}$ in the metric $g$. Next we introduce a map $B(g)\colon \mathcal{M} \longrightarrow \mathcal{V}$ defined by $w = B(g)h$ where

$$w^\ell = g^{ij}g^{k\ell}\left[\partial_i h_{jk} - \frac{1}{2}\partial_k h_{ij}\right].$$

This is related to the second contracted Bianchi identity, which asserts that if $h = \mathrm{Rc}(g)$ then $B(g)h = 0$. Note that $L(g)$, $D\mathrm{Rc}(g)$, and $B(g)$ are all linear partial differential operators of degrees 1, 2, and 1 respectively. The following result relates the geometry to the partial differential equations.

**5.2 LEMMA.** In the sequence

$$\mathcal{V} \xrightarrow{L(g)} \mathcal{M} \xrightarrow{D\mathrm{Rc}(g)} \mathcal{M} \xrightarrow{B(g)} \mathcal{V}$$

the compositions $D\mathrm{Rc}(g)L(g)$ and $B(g)D\mathrm{Rc}(g)$ all have degree 1.

<u>Proof</u>. If we apply a diffeomorphism to a metric $g$ we change its Ricci

tensor $h = Rc(g)$ by the same diffeomorphism. Hence $DRc(g)L(g) = L(h)$, which is degree 1. On the other hand, if we differentiate the second Bianchi identity $B(g)Rc(g) = 0$ we get

$$B(g)\ DRc(g)\tilde{g} + DB(g)\{Rc(g), \tilde{g}\} = 0.$$

Since $B(g)$ has degree 1 in $g$, its derivative $DB(g)\{h,\tilde{g}\}$ has degree 1 in $\tilde{g}$, so $B(g)DRc(g)\tilde{g}$ is also of degree 1 in $\tilde{g}$.

If we compose an operator of degree 2 with an operator of degree 1 we expect an operator of degree 3. The composition has lower order only when the highest derivatives cancel. This happens in this situation due to the invariance of the problem under the diffeomorphism group. It follows that the operator $DRc(g)$ cannot be elliptic. For the image of the symbol of $L(g)$ must lie in the null space of the symbol of $DRc(g)$, and the image of the symbol of $DRc(g)$ must lie in the null space of the symbol of $B(g)$. In fact the corresponding sequence of symbols

$$0 \longrightarrow V \xrightarrow{\sigma L(g)\xi} M \xrightarrow{\sigma DRc(g)\xi} M \xrightarrow{\sigma B(g)\xi} V \longrightarrow 0$$

is exact, where V and M are the vector bundles of vector fields and symmetric tensors and $\xi$ is a non-zero cotangent vector.

If in addition we define a linear operator $A(g)\colon \mathcal{M} \longrightarrow \mathcal{M}$ by $A(g)h = k$ where

$$k_{ij} = \frac{1}{2}\, g^{pq}\, h_{pq}\, g_{ij} - h_{ij}$$

then it is easy to compute that $A(g)$ is invertible with inverse $A(g)^{-1}k = h$ given by

$$h_{ij} = \frac{1}{n-2}\, g^{pq}\, k_{pq}\, g_{ij} - k_{ij}$$

when $n > 2$. In addition

$$B(g) = L(g)^{*}A(g)$$

where $L(g)^*$ is the adjoint operator of $L(g)$, and $A(g)DR(g)$ is self-adjoint. Since $A(g)$ is invertible, the sequence

$$V \xrightarrow{L(g)} \mathfrak{M} \xrightarrow{DRc(g)} \mathfrak{M} \xrightarrow{L(g)^* A(g) = B(g)} V$$

is equivalent to the sequence

$$V \xrightarrow{L(g)} \mathfrak{M} \xrightarrow{A(g)DRc(g)} \mathfrak{M} \xrightarrow{L(g)^*} V$$

and the second sequence is self-adjoint.

The fact that $A(g)DRc(g)$ is self-adjoint may be seen directly from the variational formulas in section 3. We saw that for the function

$$J_h(g) = \int Tr_g \left[ Rc(g) - h \right] \mu$$

we have

$$DJ_h(g)\tilde{g} = < Rc(g) - h,\ A(g)\tilde{g} >.$$

It follows that when $Rc(g) = h$ we have

$$D^2 J_h(g)\{\tilde{g}_1, \tilde{g}_2\} = < DRc(g)\tilde{g}_1,\ A(g)\tilde{g}_2 >.$$

But the second derivative is always symmetric, and $A(g)$ is self-adjoint. The result follows.

Our strategy for inverting $DRc(g)$, up to a null space of dimension one and an image of codimension one, is to consider the problems of the null space and image separately. Anything in the null space of $DRc(g)$ is also in the null space of $B(g)DRc(g)$ also, and this extra equation of lower order is a big help. Likewise everything in the image of $DRc(g)L(g)$ is also in the image of $DRc(g)$, and again this lower order equation is a big help.

**5.3 LEMMA.** If $g$ is near the constant curvature +1 metric $\bar{g}$ then $DRc(g)$ has a one-dimensional null space spanned by $g$ and a closed image of codimension one consisting of those tensors perpendicular to $g$.

Proof. It is automatic that $g$ lies in the null space of $DRc(g)$ because of the dilation invariance. It is also clear that the image of $DRc(g)$ is perpendicular to $g$ because of its formula as covariant derivatives which vanish under integration by parts.

To see that nothing else occurs in the null space, we examine the expanded system

$$DRc(g) \oplus B(g)DRc(g): \mathcal{M} \longrightarrow \mathcal{M} \oplus V.$$

Treating $DRc(g)$ as an operator of degree 2 and $B(g)DRc(g)$ as an operator of degree 1, this is a weighted overdetermined system. It is easily seen to be elliptic by examining the symbol to see that it is injective. It suffices to do this at the constant curvature case, since the notion is stable under small perturbations. We can compute the null space at the constant curvature case directly, to see it contains only $g$. Then the same must be true of the nearby maps.

To see that the image contains everything perpendicular to $g$, we examine the expanded system

$$DRc(g) \oplus DRc(g)L(g): \mathcal{M} \oplus V \longrightarrow \mathcal{M}.$$

Treating $DRc(g)$ as an operator of degree 2, and $DRc(g)L(g)$ as an operator of degree 1, this is a weighted underdetermined system. It is easily seen to be elliptic by examining the symbol to see that it is surjective. It suffices to check this at the constant curvature case, since the motion is stable under small perturbations. We can compute the image at the constant curvature case to see it contains everything perpendicular to $g$. Then the same must be true of the nearby maps. Moreover the first problem is essentially the adjoint of the second, if we conjugate by the invertible transformation $A(g)$. Hence it suffices to do only the second.

When $g$ has constant curvature $+1$ then $h = Rc(g) = g$ and $DRc(g)L(g) = L(h) = L(g)$. Therefore we need only consider the system

$$DRc(g) \oplus L(g) : \mathcal{M} \oplus V \longrightarrow \mathcal{M}.$$

Recall that if $DRc(g)\tilde{g} = \tilde{h}$ then

$$\tilde{h}_{ij} = \frac{1}{2} g^{k\ell} \left[\partial_i\partial_k\tilde{g}_{j\ell} + \partial_j\partial_\ell\tilde{g}_{ik} - \partial_i\partial_j\tilde{g}_{k\ell} - \partial_k\partial_\ell\tilde{g}_{ij}\right]$$

and if $L(g)\tilde{v} = \tilde{g}$ then

$$\tilde{g}_{ij} = g_{ik}\partial_j\tilde{v}^k + g_{jk}\partial_i\tilde{v}^k.$$

If we put $\tilde{v} = D(g)\tilde{g}$ where

$$\tilde{v}^\ell = \frac{1}{2} g^{ij}g^{k\ell}\partial_i\tilde{g}_{jk}$$

then

$$L(g)D(g)\tilde{g} - DRc(g)\tilde{g} = E(g)\tilde{g}$$

where $E(g)\tilde{g} = \tilde{p}$ is given by

$$\tilde{p}_{ij} = \frac{1}{2} g^{k\ell} \left[\partial_i\partial_j\tilde{g}_{k\ell} + \partial_k\partial_\ell\tilde{g}_{ij}\right].$$

Therefore an inverse for $E(g)$ will provide a one-sided inverse for $DRc(g) \oplus L(g)$. The same holds at the symbol level. The symbol of $E(g)$ in the direction $\xi$ is

$$\tilde{p}_{ij} = \frac{1}{2} g^{k\ell} \left[\xi_i\xi_j\,\tilde{g}_{k\ell} + \xi_k\xi_\ell\,\tilde{g}_{ij}\right]$$

and this is one-to-one if and only if it is onto. If

$$g^{k\ell}\left[\xi_i\xi_j\,\tilde{g}_{k\ell} + \xi_k\xi_\ell\tilde{g}_{ij}\right] = 0$$

then contracting with $g^{ij}$ we get $g^{ij}\tilde{g}_{ij} = 0$ if $g^{k\ell}\xi_k\xi_\ell \neq 0$, and hence $\tilde{g}_{ij} = 0$. This shows E(g) is elliptic, and hence DR(g) $\oplus$ L(g) is an underdetermined weighted elliptic system. If $E(g)\tilde{g} = 0$ then

$$g^{k\ell}\left[\partial_i\partial_j\tilde{g}_{k\ell} + \partial_k\partial_\ell\,\tilde{g}_{ij}\right] = 0.$$

Contracting with $g^{ij}$ we have

$$\Delta(g^{ij}\,\tilde{g}_{ij}) = 0$$

so $g^{ij}\tilde{g}_{ij}$ is constant. Then $\Delta\tilde{g}_{ij} = 0$. Integrating by parts we get

$$\int g^{i\ell}g^{jm}g^{kn}\partial_i\tilde{g}_{jk}\partial_\ell\tilde{g}_{mn} = 0$$

so $\partial_i\tilde{g}_{jk} = 0$. When $g_{ij}$ has constant curvature +1 then for any $\tilde{g}_{ij}$,

$$g^{ik}\left[\partial_i\partial_j\tilde{g}_{k\ell} - \partial_j\partial_i\tilde{g}_{k\ell}\right] = n\,\tilde{g}_{j\ell} - g^{ik}\tilde{g}_{ik}g_{j\ell}.$$

If in addition $\partial_i\tilde{g}_{jk} = 0$ then $g^{ik}\tilde{g}_{ik}$ is a constant, and $\tilde{g}_{j\ell}$ is a constant multiple of $g_{j\ell}$. Hence the null space of E(g) is one dimensional. We can deform E(g) through elliptic operators to the self-adjoint operator $\frac{1}{2}\,g^{k\ell}\partial_k\partial_\ell\tilde{g}_{ij}$, so E(g) has index zero. Therefore its image is closed and has codimension 1. The image of DRc(g) $\oplus$ L(g) contains the image of E(g), so it is also closed of codimension 1. This finishes the proof of Lemma 5.3.

We can now prove Theorem 5.1. We define a map

$$P: \mathcal{M}^+ \times R^+ \subset \mathcal{M} \times R \longrightarrow \mathcal{M} \times R$$

by letting

$$P(g,c) = (c\ Rc(g),\ V(g)).$$

We claim P is locally invertible near $(\bar{g},1)$ when $\bar{g}$ is the constant curvature metric. Because of our characterization of DRc(g) it follows immediately that DP(g,c) is invertible for all (g,c) near $(\bar{g},1)$. Moreover the inverse may be constructed out of Green's operator for the elliptic operator

$$E(g) = DRc(g)\ L(g)\ D(g) - DRc(g).$$

It follows from the results of [2] on elliptic operators that all the maps constructed belong to the Nash-Moser category. Therefore P is locally invertible. This proves Theorem 5.1.

## 6. Equivariant Examples

We compute two special cases for metrics with a large group of symmetries. In the first case we regard $S^3$ as the Lie group of unit quaternions acting on itself by left multiplication. Any left-invariant metric will have a left-invariant Ricci tensor, which we will compute. The Lie bracket uniquely determines a reference metric of constant curvature +1 such that for any two left-invariant vector fields X and Y the Lie bracket is twice the cross-product

$$[X,Y] = 2X\times Y.$$

By diagonalizing a given left-invariant metric g with respect to this reference we obtain a basis U, V, W such that

$$[U,V] = 2W \qquad [V,W] = 2U \qquad [W,U] = 2V$$

and

$$g(U,U) = u \qquad g(V,V) = v \qquad g(W,W) = w$$

for some constants u, v, w while

$$g(U,V) = 0 \qquad g(V,W) = 0 \qquad g(W,U) = 0.$$

We can compute the Levi-Civita connection of the metric g using the fact that for any left-invariant vector fields X, Y,Z we have

$$g(X \cdot Y,Z) = \frac{1}{2} \{g([X,Y],Z) - g([X,Z],Y) - g([Y,Z],X)\},$$

where $X \cdot Y$ is the covariant derivative in the direction X of Y. Then we can compute

$$U \cdot V = \frac{x}{w} W \qquad V \cdot W = \frac{y}{u} U \qquad W \cdot U = \frac{z}{v} V$$

$$V \cdot U = -\frac{y}{w} W \qquad W \cdot V = -\frac{z}{u} U \qquad U \cdot W = -\frac{x}{v} V$$

where

$$x = v + w - u \qquad y = u + w - v \qquad z = u + v - w$$

and the other covariant derivatives are zero. The curvature of the connection is given by

$$R(X,Y)Z = X \cdot (Y \cdot Z) - Y \cdot (X \cdot Z) - [X,Y] \cdot Z$$

$$R(X,Y,X,Y) = g\ (R(X,Y)Y,X)$$

and we compute

$$R(U,V,U,V) = 2 \left\{z - \frac{xy}{x+y}\right\}$$

$$R(V,W,V,W) = 2 \left\{x - \frac{yz}{y+z}\right\}$$

$$R(W,U,W,U) = 2 \left\{y - \frac{xz}{x+z}\right\}.$$

We can then compute the Ricci tensor as

$$Rc(U,U) = \frac{yz}{vw}$$

$$Rc(V,V) = \frac{xz}{uw}$$

$$Rc(W,W) = \frac{xy}{uv},$$

while the other terms are zero.

**6.1 THEOREM.** For every equivariant tensor $h > 0$ there exists a unique constant $c > 0$ such that $Rc(g) = ch$ for some equivariant $g > 0$, and $g$ is unique up to a constant.

<u>Proof</u>. Since g,h and the reference tensor will all be simultaneously diagonal, knowing h determines the basis U, V, W. Let

$$h(U,U) = r \qquad h(V,V) = s \qquad h(W,W) = t.$$

We must solve for $u, v, w > 0$ such that if

$$x = v+w-u \qquad y = u+w-v \qquad z = u+v-w$$

then

$$\frac{yz}{vw} = cr \qquad \frac{xz}{uw} = cs \qquad \frac{xy}{uv} = ct.$$

Given $x, y, z > 0$ we can recover $u, v, w > 0$ by the formulas

$$u = \frac{y+z}{2} \qquad v = \frac{x+z}{2} \qquad w = \frac{x+y}{2}.$$

Since g is only determined up to a constant, we may as well assume that

$$cuvw = 1.$$

Then our equations become

$$yzu = r \qquad xzv = s \qquad xyw = t,$$

which we can also write as

$$yz(y+z) = 2r \qquad xz(x+z) = 2s \qquad xy(x+y) = 2t.$$

We claim that for every r,s,t, > 0 there exists a unique solution x,y,z > 0 for these equations. We can then recover u, v, w > 0 and c > 0 as above.

To see this, substitute

$$p = xyz \qquad q = p(x+y+z).$$

Then we get the equations

$$x = \frac{q}{p+2r} \qquad y = \frac{q}{p+2s} \qquad z = \frac{q}{p+2t}.$$

Substituting this in $q = p(x+y+z)$ we get

$$\frac{p}{p+2r} + \frac{p}{p+2s} + \frac{p}{p+2t} = 1.$$

Now each term is monotone increasing in p as p goes from zero to infinity, with the term going from zero to one. Hence there is exactly one value of $p > 0$ solving the equation for any given r, s, t > 0. Since $xyz = p$ we have

$$q^3 = p(p+2r)(p+2s)(p+2t)$$

which gives a unique value for $q > 0$. We then recover x, y, z > 0 from the formulas above. This proves the theorem.

For our second example we consider metrics on $S^3 \subseteq R^4 = R^3 \times R^1$ invariant under the rotation group $0(3) \times 0(1)$. Let $R = \{(t,u,v,w)\}$ and parametrize $S^3 = \{t^2 + u^2 + v^2 + w^2 = 1\}$ by angles x, y, z setting

$$t = \cos x \cos y \cos z$$
$$u = \cos x \cos y \sin z$$
$$v = \cos x \sin y$$
$$w = \sin x$$

as the parametrization. Under the group action of 0(3) the fibres are spheres $S^2$ except at the ends where they are points, and the orbit space is an interval. The reflection 0(1) folds the interval about the middle.

We regard x as the coordinate along the orbit space and y, z as the coordinates on the sphere fibres. A metric tensor $ds^2 = g_{ij}dx^i dx^j$ invariant under the group will have a distance f(x) dx between orbits, and if the orbit at x is a sphere of radius g(x) it will have a rotationally invariant metric $g(x)^2(dy^2 + \cos^2 y\, dz^2)$ along the orbit, while the slices $\{x = \text{constant}\}$ and $\{y,z = \text{constant}\}$ must be perpendicular since we can rotate $S^2$ around any point by 180°. Therefore an equivariant metric tensor will have the form

$$ds^2 = f(x)^2\, dx^2 + g(x)^2(dy^2 + \cos^2 y\, dz^2)$$

where f(x) and g(x) are smooth functions of x on $0 \leqslant x \leqslant \pi/2$ and

(a) f(x) and g(x) are even at $x = 0$

(b) f(x) is even and g(x) is odd at $x = \pi/2$

(c) $f(\pi/2) + g'(\pi/2) = 0$

(d) $f(x) > 0$ on the whole interval and
$g(x) > 0$ except at the end point.

Conversely these conditions imply that the metric is smooth and positive. To check the endpoint condition at $x = \pi/2$, let $\rho = \cos x$. Then

$$t = \rho \cos y \cos z$$

$$u = \rho \cos y \sin z$$

$$v = \rho \sin y$$

describes spherical coordinates on t, u, v space, while t, u, v are a coordinate chart on $S^3$ at the point $t = u = v = 0$, $w = 1$. A smooth rotationally invariant metric in t, u, v space near the origin must be a linear combination of $ds^2$ and $d\rho^2$ and so has the form

$$ds^2 = \phi(t,u,v)(dt^2 + du^2 + dv^2) + \psi(t,u,v)(t\,dt + u\,du + v\,dv)^2$$

where $\phi$ and $\psi$ are smooth rotationally invariant functions near zero with $\phi(0) > 0$. In terms of $\rho$, y, z we must have $\phi(t,u,v) = \phi(\rho^2)$ and $\psi(t,u,v) = \phi(\rho^2)$ two smooth functions of $\rho^2$, and

$$ds^2 = \phi(\rho^2)[d\rho^2 + \rho^2(dy^2 + \cos^2 y\,dz^2)] + \psi(\rho^2)\,\rho^2\,d\rho^2.$$

Near $x = \pi/2$ the function $\rho = \cos x$ is a smooth invertible function of x which is odd in x around $\pi/2$, so smoothness in x is equivalent to smoothness in $\rho$, and oddness or eveness in x around $\pi/2$ is the same as oddness or evenness in $\rho$ around 0. Then

$$ds^2 = f(\rho)^2\,d\rho^2 + g(\rho)^2(dy^2 + \cos^2 y\,dz^2)$$

where

$$f(\rho)^2 = \phi(\rho^2) + \rho^2\psi(\rho^2) \qquad g(\rho)^2 = \rho^2\phi(\rho^2).$$

It follows that $f(\rho)$ and $g(\rho)$ are smooth functions of $\rho$ and that at $\rho = 0$ we have $f(\rho)$ even in $\rho$ and $g(\rho)$ odd in $\rho$. Moreover

$$\frac{dg}{d\rho}(0) = \lim_{\rho\to 0} \frac{g(\rho)}{\rho} = \sqrt{\phi(0)} = f(0).$$

Then since $d\rho/dx = -1$ at $x = 0$, we have

$$\frac{dg}{dx}(0) = - f(0).$$

This proves the validity of the end point conditions. Conversely, given $f(\rho)$ and $g(\rho)$ satisfying the conditions, we recover $\phi(\rho^2) = g(\rho^2)/\rho^2$ as a smooth function with $\phi(0) > 0$, and we also recover $\psi(\rho^2) = [f(\rho)^2 - \phi(\rho^2)]/\rho^2$ as a smooth function. This requires two elementary facts; a smooth even function of $\rho$ is a smooth function of $\rho^2$, and if a smooth function of $\rho^2$ vanishes at $\rho = 0$ then its quotient by $\rho^2$ is smooth.

We now compute the curvature of the metric

$$ds^2 = f(x)^2\, dx^2 + g(x)^2[dy^2 + \cos^2 y\, dz^2].$$

The non-vanishing Christoffel symbols are

$$\Gamma^x_{xx} = \frac{f'}{f} \qquad \Gamma^x_{yy} = -\frac{gg'}{f^2} \qquad \Gamma^x_{zz} = -\frac{gg'}{f^2}\cos^2 y$$

$$\Gamma^y_{xy} = \frac{g'}{g} \qquad \Gamma^y_{zz} = \sin y \cos y$$

$$\Gamma^z_{xz} = \frac{g'}{g} \qquad \Gamma^z_{yz} = - \tan y.$$

The Riemannian curvature is

$$R_{xyxy} = - fg \left(\frac{g'}{f}\right)'$$

$$R_{xzxz} = - fg \left(\frac{g'}{f}\right)' \cos^2 y$$

$$R_{yzyz} = g^2\left[1 - \left(\frac{g'}{f}\right)^2\right] \cos^2 y$$

while the other independent terms are zero. The Ricci curvature is

$$R_{xx} = - \frac{f}{g}\left(\frac{g'}{f}\right)'$$

$$R_{yy} = \frac{1}{2}\left[1 - \left(\frac{g'}{f}\right)^2 - \frac{g}{f}\left(\frac{g'}{f}\right)'\right]$$

$$R_{zz} = R_{yy} \cos^2 y$$

while the other terms are zero. If we let $d\sigma^2 = R_{ij}dx^i dx^j$ then the Ricci metric $d\sigma^2$ is given by

$$d\sigma^2 = h(x)^2 dx^2 + k(x)^2 [dy^2 + \cos^2 y \; dz^2]$$

where

$$h^2 = -\frac{f}{g}\left(\frac{g'}{f}\right)' \qquad k^2 = \frac{1}{2}\left[1 - \left(\frac{g'}{f}\right)^2 - \frac{g}{f}\left(\frac{g'}{f}\right)'\right].$$

**6.2 LEMMA.** The metric

$$ds^2 = f(x)^2 \, dx^2 + g(x)^2 [dy^2 + \cos^2 y \; dz^2]$$

has positive Ricci curvature if and only if

$$\left(\frac{g'}{f}\right)' < 0$$

in the interval $0 \leqslant x < \pi/2$, and vanishes but with non-zero derivative at the end point $x = \pi/2$.

<u>**Proof**</u>. The condition is necessary to make $h > 0$. If it holds then $g'/f$ decreases from zero at $x = 0$, to $-1$ at $x = \pi/2$. Then $1 - (g'/f)^2 > 0$ on the interval and $k > 0$ also. We can likewise check the endpoint.

**6.3 THEOREM.** If two metrics on $S^3$ equivariant under $0(3) \times 0(1)$ have the same Ricci tensor they differ by a constant multiple. If $h > 0$ is a metric of positive Ricci curvature on $S^3$ equivariant under $0(3) \times 0(1)$, there exists a unique positive constant c such that $Rc(g) = ch$ for some equivariant metric g, which is unique up to a constant. However, there exist metrics $h > 0$ which do not have positive Ricci curvature such that no positive multiple ch is the Ricci tensor of any equivariant metric.

**Proof.** The metric g becomes $ds^2 = f^2\, dx^2 + g^2(dy^2 + \cos^2 y\, dz^2)$ and the metric h becomes $d\sigma^2 = h^2\, dx^2 + k^2(dy^2 + \cos^2 y\, dz^2)$.

We make the substitution $\ell = -g'/f$. Then the Ricci curvature equations become

$$h^2 = \frac{f}{g}\,\ell' \qquad k^2 = \frac{1}{2}\left[1 - \ell^2 + \frac{g}{f}\,\ell'\right].$$

Solving the first for g/f and substituting in the second we soon find that

$$\ell' = h(2k^2 + \ell^2 - 1)^{1/2}.$$

Since we know that $\ell = 0$ at $x = 0$, it follows that $\ell$ is uniquely determined by h and k. We can then solve for g up to a constant from the equation

$$g'/g = -\,h^2\ell/\ell'$$

and solve for $f = -\,g'/\ell$. This proves the uniqueness assertion.

To prove the existence assertion, note that by changing the x-coordinate we can assume h(x) is constant. Since we are allowed to dilate by a constant, we take $h(x) = 1$. Then multiplying h and k by a constant c, we get the equations

$$\frac{f}{g}\,\ell' = c^2 \qquad\qquad \frac{1}{2}\left[1 - \ell^2 + \frac{g}{f}\,\ell'\right] = c^2k^2$$

which reduce as before to the equation

$$\ell' = c\left[(2c^2k^2 + \ell^2 - 1)\right]^{1/2}.$$

We must solve this equation with $\ell = 0$ at $x = 0$ by the endpoint conditions. The solution will exist on all of $0 \leqslant x \leqslant \pi/2$ provided c is large enough. But by the endpoint conditions we must also arrange to have $\ell = 1$ at $x = \pi/2$, and this will determine the constant c. The hypothesis that the metric $d\sigma^2 = h^2\, dx^2 + k^2(dy^2 +$

$\cos^2 y\ dx^2$) itself has positive Ricci curvature is equivalent by Lemma 6.2 to the assertion that $(k'/h)' < 0$, and since we normalized to $h = 1$ this is just the assertion that $k'' < 0$, or that $k$ is concave. Since $k' = 0$ at $x = 0$ and $k = 0$ at $x = \pi/2$ by the symmetry conditions, this implies that $k$ is monotone decreasing. The procedure for solving the equation is to pick a large $c$ for which the solution exists on the interval, and to decrease $c$ until $2c^2k^2 + \ell^2 - 1 = 0$ at some point. If this happens at the endpoint $x = \pi/2$ where $k = 0$, then $\ell = 1$ there and we are done. If it happens at an interior point, then the derivative of $2\ c^2k^2 + \ell^2 - 1$ vanishes there also, so

$$4c^2kk' + 2\ell\ell' = 0$$

at that point. But $\ell' = 0$ at the point, so $kk' = 0$ at the point. Our hypothesis on $k$ makes $k > 0$ and $k' < 0$ at interior points. Hence there exists a solution.

If we drop the convexity hypothesis it is easy to find an example of a function $k$ for which there is no solution for any $c$. Just make $k(0)$ small but let $k(x)$ get very large in the interior. Since $k(0)$ is small we must make $c$ large for the solution to start. Then when $k$ gets large $\ell$ will get bigger than 1, and then remain so. Thus we cannot get $\ell = 1$ at $x = \pi/2$. This proves the theorem.

Note that this example strongly suggests that to solve $Rc(g) = ch$ for some $c$, we must impose some restriction on $h$. Since $h$ is just a metric tensor and $g$ is still unknown, the only reasonable condition on $h$ must be some metric invariant, such as the curvature. And this is just the condition we see in the theorem.

We conclude with the following remark. If $Rc(g\ ) = h$ then $h$ lies along some submanifold which we expect to have codimension 1. It is natural to ask what equation $h$ must satisfy to be a Ricci tensor. In dimension two it is a restriction on the Ricci volume, the volume of the Ricci metric $h$, given by the Gauss-Bonnet formula. It is natural therefore to ask if there is any restriction on the Ricci volume in higher dimensions. The answer is no.

**6.4 THEOREM.** There exist metrics on $S^3$ of positive Ricci curvature whose Ricci volume is any number between 0 and $\infty$.

<u>Proof</u>. For small Ricci volume we consider a metric on $S^3$ of the first kind, invariant under $S^3$. We can pick $x, y, z > 0$ arbitrarily. Then the Ricci volume is

$$V = \frac{xyz}{uvw} = \frac{8xyz}{(x+y)(x+z)(y+z)}.$$

Take $y = 1$ and $z = 1$. Then

$$V = \frac{4x}{(x+1)^2}$$

which can be any number between 0 and 1.

For large Ricci volume we consider a metric of the second kind

$$ds^2 = f(x)^2\,dx^2 + g(x)^2(dy^2 + \cos^2 y\,dz^2)$$

invariant under $0(3) \times 0(1)$. Its volume is

$$V = 8\pi \int_0^1 f(x)\,g(x)^2 dx.$$

The Ricci tensor

$$d\sigma^2 = h(x)^2 dx^2 + k(x)^2(dy^2 + \cos^2 y\,dz^2)$$

has Ricci volume

$$V = 8\pi \int_0^1 h(x)\,k(x)^2\,dx.$$

Taking $f = 1$ we get the Ricci volume

$$V = 4\pi \int_0^1 \left(- \frac{g''}{g}\right)^{1/2} \left[1 - (g')^2 - gg''\right] dx.$$

For any $\epsilon > 0$ we choose a smooth function $g(x)$ which will have $g(x) \approx \pi/2 - x$ on $\epsilon \leqslant x \leqslant \pi/2$ and $g''(x) \approx - 1/\epsilon$ on $0 \leqslant x \leqslant \epsilon$. We can then make $g(x)$ satisfy all the conditions for a smooth metric and in addition have $0 \leqslant g' \leqslant 1$, while being close to the function above. This makes

$$V \geqslant 4\pi \int_0^1 \left(\frac{(-g'')^3}{g}\right)^{1/2} dx.$$

Note we keep $g'' < 0$ for positive Ricci curvature. Since $g \approx 1$ and $-g'' \approx 1/\epsilon$ on $0 \vee x \leqslant \epsilon$ we get $V \approx \epsilon^{-1/2}$, which can be arbitrarily large. This proves the result.

## References


[1] DeTurck, Dennis, Metrics wih prescribed Ricci curvature, Seminar on Differential Geometry, Princeton U. Press, 1982, pp. 706.

[2] Hamilton, R.S., The inverse function theorem of Nash and Moser, Bull. A.M.S., Vol. 7, No.1, 1982, pp. 65-222.

[3] Hamilton, R.S., Three-manifolds with positive Ricci curvature, J. Diff. Geom., Vol. 17, 1982, pp. 255-306.

# A Walk Through Partial Differential Equations

By Fritz John

This is an informal talk for non-specialists dealing on an elementary level with some aspects of the theory of partial differential equations. In recent years progress in the theory has been tremendous, often in unexpected directions, while also solving classical problems in more general settings. New fields have been added, like the study of variational inequalities, of solitons, of wave front sets, of pseudo-differential operators, of differential forms on manifolds, etc. Much of the progress has been made possible by the use of functional analysis. However, in the process much of the original simplicity of the theory has been lost. This is perhaps connected with the emphasis on solving problems, which often requires the piling up mountains of a priori inequalities and the skillful juggling of function spaces to make ends meet. It is good to remember that mathematics is not only concerned with solving problems, but with studying the structure and behavior of the objects it creates. One of the best examples is the classical theory of functions of a complex variable. It, incidentally, does solve problems as in the Riemann mapping theorem. But much of its beauty lies in statements that can hardly be considered as "solving" anything, like the calculus of residues, or Picard's theorem, or Cauchy's formula

$$(1) \qquad f(z) = \frac{1}{2\pi i} \int_C \frac{f(\zeta)}{\zeta - z}\, d\zeta$$

The only "problem" solved by (1) is the improper one of determining f from its values on C, which generally has no solution. But that is not the reason for its importance. Formula (1) not only is strikingly beautiful but extremely useful. It shows immediately that analytic f can be differentiated infinitely often and can be represented by convergent power series.

The theory of complex analytic functions to some extent can serve as a model for that of solutions of linear partial differential equations. Of course, complex function theory is just the theory of the special systems of Cauchy-Riemann equations. It is natural to generalize the theory to other equations and systems. One possibility is to stick closely to some analogues of the classical Cauchy-Riemann equations as is done, for instance, in the theory of functions of several complex variables or in that of pseudo-analytic functions (see [1], [2]) or in that of systems associated with hypercomplex numbers (see [3]) or in potential theory in more dimensions. Quite generally, however, solutions of any linear elliptic system of partial differential equations have much in common with analytic functions. This can be seen by rather elementary arguments. I restrict myself here to linear equations whose coefficients are either in $C^\infty$ or are in $C^a$ (i.e., are real analytic, locally represented by power series for real arguments, their extensions to complex arguments will not be used).

The following well-known properties of analytic f(z) have their analogues for solutions of elliptic systems:

a) f(z) inside a domain can be represented by a boundary integral, as in (1)

b) If f as a function of $z = x + iy$ is in $C^1$, then it is in $C^\infty$.

c) An isolated singularity of f(z) at a point $\zeta$ is either

  ($\alpha$) removable

  ($\beta$) of finite integer order ("pole")

  ($\gamma$) of infinite order ("essential")

  Correspondingly, we have a Laurent expansion

$$f(z) = \sum_{k=-\infty}^{+\infty} c_k (z-\zeta)^k$$

  with either no terms with $k < 0$, or a finite or infinite number of such terms.

d) Continuation of f from open sets into larger connected sets is unique.

Some of the analogues of these statements for solutions of elliptic equations follow directly by calculus. We consider linear m-th order equations

(2) $$Lu = \sum_{|\alpha| \leqslant m} A_\alpha(x) D^\alpha u = 0$$

for a scalar $u = u(x_1, x_2, \dots, x_n) = u(x.)$ (Everything that follows can be extended to systems of such equations, if we interpret u in (2) as an N-vector and the coefficients $A_\alpha$ as NxN-matrices.) We first look at the question of regularity of solutions. (For details see [4]). (2) represents one linear relation between the $D^\alpha u$ with $|\alpha| \leqslant m$. Things become interesting if we consider (2) along a hypersurface, say a sphere $S_{z,r}$ with center z and radius r, where other relations ("consistency conditions") are satisfied by the $D^\alpha u$. In each point of $S_{z,r}$ we have a unit normal represented by the vector $\xi = (x-z)/r$. Every derivative in a point of $S_{z,r}$ can be decomposed into a normal, a tangential, and a lower order derivative. More precisely, an N-th order operator $\Sigma B_\beta D^\beta$ can be represented in the form

(3) $$\sum_{|\beta| \leqslant N} B_\beta D^\beta = \left[\sum_{|\beta| = N} B_\beta \xi^\beta\right] \left[\sum_k \xi_k D_k\right]^N + \sum_{\substack{k \\ |\beta| = N-1}} a_{k\beta} D_k D^\beta + \sum_{|\beta| \leqslant N-1} d_\beta D^\beta$$

where the $a_k = a_k\,(x,z)$ satisfy the tangentiality condition

(4) $$\sum_k \xi_k a_{k\beta} = 0$$

The differential equation (2) permits to eliminate the normal derivative term in (3) for $N \geqslant m$, provided the operator L is elliptic in the sense that

(5) $$\sum_{|\alpha| = m} A_\alpha \xi^\alpha \neq 0 \text{ for any } \xi \neq 0$$

Spherical integrals of tangential derivatives can be reduced to integrals of lower order derivatives by integration by parts:

$$\int_{S_{z,r}} \sum_k a_{k\beta} (D_k\theta)\, dS = - \int_{S_{z,r}} \left[\sum_k D_k a_{k\beta}\right] \theta\, dS$$

if (4) holds. Altogether we find that for solutions u of (2) there exists an identity

$$\int_{S_{z,r}} \sum_\beta b_\beta D^\beta u\, dS = \int_{S_{z,r}} \sum_{|\beta| \leq m-1} c_\beta D^\beta u\, dS$$

with coefficients $c_\beta$ depending on x,z,r. It follows immediately that for a solution u of (2) the integral of u over a sphere $S_{z,r}$ is of class $C^\infty$ in z and r, provided $A_\alpha \in C^\infty$, $u \in C^m$, $r>0$.

It remains to show that a function u(x) with spherical integrals in $C^\infty$, itself, belongs to $C^\infty$. We show this just for n=3 using Poisson's expression for the solution of the initial value problem for the wave equation in terms of spherical means. According to this the solution v(x,t) of

(6a) $\quad v_{tt} - \Delta v = 0$

(6b) $\quad v(x,0) = u(x),\ v_t(x,0) = 0$

at the point (z,t) is given by

$$v(z,t) = \Omega_t u = \frac{1}{4\pi t^2} \int_{S_{z,t}} \left[u(x) + t \sum_k \xi_k D_k u(x)\right] dS$$

Thus $v(z,t) \in C^\infty$ for $t>0$ for our solution u of (2). Now the solution v of (6a,b) can be written symbolically as

$$v = \Omega_t u = \cos(\sqrt{-\Delta}\ t)\ u$$

Using the identity

$$1 = 2\cos^2\theta - \cos(2\theta)$$

we find an expression or u in terms of v(z,t) for t>0:

$$u(z) = 2(\Omega_t)^2 u - \Omega_{2t}\, u = 2\Omega_t v(z,t) - v(z,2t)$$

where t is any positive number. It follows immediately that $u \in C^\infty$ if $v \in C^\infty$ and, hence, that solutions of an elliptic linear equation with coefficients in $C^\infty$ are themselves in $C^\infty$ if in $C^m$. The same arguments yield that $u \in C^a$, if the $A_\alpha$ are in $C^a$, even if u is only a continuous weak solution of (2).

The preceding proof of regularity of solutions uses nothing but purely formal algebraic consequences of the differential equation, in addition to the ellipticity condition (5). In order to derive further analogies to analytic functions we have to resort to the Lagrange-Green identity

$$(7) \qquad \int_\omega [v(Lu) - (\bar{L}v)u]\, dx = \int_{\partial\omega} M(u,v,\xi)\, dS$$

where $\bar{L}$ is the differential operator formally adjoint to L, and M is bilinear in the derivatives of u and v of orders $\leq m-1$, and is linear in the components of the unit normal $\xi$ of $\partial\omega$. We get an analogue to Cauchy's formula (1), if we take here for v a <u>fundamental solution</u> K(x,z) for which

$$(8) \qquad \bar{L}K(x,z) = \delta(x-z)$$

where $\delta$ denotes the Dirac function. When Lu=0 we find from (7) the integral representation

$$(9) \qquad u(z) = -\int_{\partial\omega} M(u,K,\xi)\, dS$$

Thus, K here plays the role of the kernel $1/(\zeta-z)$ in (1). It is a solution of $\bar{L}K=0$ with a singularity at z, which is isolated in the elliptic case. Such solutions are easily constructed when L has real analytic coefficients $A_\alpha$. (See [4], [5], [6].) For simplicity we again restrict ourselves to the case n=3. We notice that by potential theory $r=|x-z|$ satisfies

$$\delta(x-z) = -\frac{1}{4\pi}\,\Delta\,\frac{1}{r} = -\frac{1}{8\pi}\,\Delta^2\, r$$

$$= -\frac{1}{8\pi^2}\,\Delta^2 \int_{|\xi|=1} H((x-z)\cdot\xi)\, d\omega_\xi$$

where H is the Heaviside function. Let $k(x,z,\xi)$ denote the solution of the Cauchy problem

$$\bar{L}k = (x-z)\cdot\xi$$

$$D^\alpha k = 0 \text{ for } |\alpha| \leqslant m-1,\ (x-z)\cdot\xi = 0$$

The Cauchy-Kowalevski theorem guarantees the existence of k for x near z and $|\xi| = 1$ (the plane $(x-z)\cdot\xi = 0$ is non-characteristic because L is elliptic). Then

$$(10) \qquad K(x,z) = -\frac{1}{8\pi^2}\,\Delta^2 \int_{|\xi|=1} H(k(x,z,\xi))\, d\omega_\xi$$

is a fundamental solution satisfying (8). One easily verifies that K(x,z) is real analytic in x,z for x near z and $x \neq z$. The analyticity of u(z) then follows immediately from (9).

(If the $A_\alpha$ are only in $C^\infty$, we can still construct a fundamental solution locally from a suitable integral equation ("parametrix method"; see [7], [8]) and conclude that all solutions of (2) are in $C^\infty$.

In the case where the $A_\alpha$ are real analytic we can classify isolated singularities just as for complex f(z). An isolated singularity of a solution v(x) of $\bar{L}v = 0$ at a point z is called a <u>pole</u> if v and its derivatives become $\infty$ at z at most like a reciprocal power of

$r = |x-z|$. It is convenient to define the "order" s of the pole as the smallest real number such that

$$D^{\alpha}v(x) = 0(r^{-s}) \text{ for } |\alpha| = m-1.$$

One finds that the order is always an integer and that poles of the lowest possible order $s = n-1$ are represented by the fundamental solution $K(x,z)$. For example, for $n = 3$ the solution $K = -1/(4\pi r)$ of $\Delta K = \delta(x-z)$ has a pole of order 2 at z. Isolated singularities of order < n-1 are removable.

There are analogues to Laurent series in the case $A_{\alpha} \in C^{a}$. In function theory we get poles of higher order by taking powers of the "fundamental solution" $1/(\zeta - z)$. We could instead take derivatives with respect to z, and it is this feature that generalizes to solutions of elliptic equations. We can always construct a fundamental solution $K(x,z)$ which as a function of z satisfies $LK = 0$ for $z \neq x$. Let $v(z)$ be a solution of $L_z v = 0$ with an isolated singularity at $z = x$, where $L_z$ indicates that coefficients and differentiations refer to z as independent variable. Then Balch [9] shows that v has a local representation

$$v(z) = w(z) + \sum_{\alpha} c_{\alpha} D_z^{\alpha} K(x,z)$$

where w is a solution of $L_z w = 0$ regular at x. The series is finite in case of a pole. The coefficients $c_{\alpha}$ in the expansion have an integral representation similar to that for complex analytic functions $f(z)$.

Finally, we discuss the unique continuation property for solutions of linear equations with coefficients $A_{\alpha} \in C^{a}$. (For the case $A_{\alpha} \in C^{\infty}$ see [8]). We have seen that in the elliptic case the solutions u themselves are in $C^{a}$ in the interior of their domain; and, hence, continue like real analytic functions: u is determined in every connected set in its domain by the values in any open subset. However, continuation can also be unique out of boundary regions (where u need not be analytic) even for non-elliptic equations. Essentially, one uses the fact that certain integral transforms of u are analytic, even if u itself is not. The fundamental insight is provided by the generalized uniqueness theorem of Holmgren derived from a combination of the Lagrange-Green identity (7) and the Cauchy-Kowalevski theorem. (See [10], [11].) Basically, one wants to show that prescribing Cauchy data of a solution u of (2) on a hypersurface S uniquely determines u in a set R adjacent to S. Let, say, $D^{\alpha}u = 0$ for $|\alpha| \leq m-1$ on S. We intersect S with a non-characteristic analytic hypersurface $\Sigma$ so that S and $\Sigma$ between them bound an n-dimensional region R. We take an analytic w and the analytic solution v of $\bar{L}v = w$ that has vanishing Cauchy data on $\Sigma$. If w exists throughout R, we find from (7) that

$$\int_R wu\, dx = 0. \tag{11}$$

Existence of v in R follows by Cauchy-Kowalevski, if R is "sufficiently" small, and that even uniformly in w for all polynomials w. The completeness of the set of polynomials then implies by (11) that u = 0 in R. One can then continue u further, starting with the initial surface $\Sigma$ instead of S with vanishing Cauchy data for u on $\Sigma$ and iterate the procedure. One is led to consider a whole family of non-characteristic analytic hypersurfaces $\Sigma_\lambda$ depending on a parameter $\lambda$ with $0 < \lambda < 1$. Each $\Sigma_\lambda$ together with S shall bound a region $R_\lambda$, where $R_\lambda$ shrinks into a point on S for $\lambda \to 0$. Then u is determined uniquely in the union of the $R_\lambda$ by its Cauchy data on S.

In the elliptic case, where all hypersurfaces are non-characteristic, one concludes that u is determined in any connected set R by its Cauchy-data on any (relatively open) portion of $\partial R$. This shows the impossibility of generally solving the Cauchy-problem for elliptic equations. Data on any portion of the boundary already uniquely determined the data over the whole boundary.

As a consequence of the generalized Holmgren theorem solutions of non-elliptic equations also an enjoy unique continuation properties, but there uniqueness may be confined to a definite portion ("domain of determinacy") of the domain of the solution, determined by the geometry of the characteristic surfaces. As an example, consider the wave equation

(12) $u_{tt} = u_{xx} + u_{yy}$

in two space dimensions. The equation is hyperbolic, and the Cauchy problem in which u and $u_t$ are prescribed for $t = 0$ is well posed. Thus prescribing u and $u_t$ on the disk

(13) $x^2 + y^2 < a^2,\ t = 0$

uniquely determines u in the double cone

(14) $\sqrt{x^2 + y^2} + |t| < a$

This can be seen from Holmgren's theorem by taking for the $\Sigma_\lambda$ suitable space-like hyperboloids filling the cone. One easily sees from examples of plane wave solutions that the Cauchy data on the set (13) do not determine u in any point outside the set (14). Here unique continuation is strictly confined to the set (14).

The situation is quite different when we prescribe Cauchy data for (12) on a time-like manifold where the data themselves have unique continuation properties and cannot be prescribed arbitrarily. Thus prescribing u and $u_x$ in a strip

(15) $x = 0,\ |y| < \varepsilon,\ |t| < a$

of arbitrary "thinness" $\varepsilon$ again determines u in the double cone (14) and, hence, determines the Cauchy data in the square

(16) $x = 0,\ |y| + |t| < a.$

Since Equation (12) is invariant under translations in the t-direction, we find that knowing the values of a solution u of (12) in any cylinder

(17) $x^2 + y^2 < \varepsilon^2, \; -\infty < t < \infty$

uniquely determines u in the whole xyt-space. It is as if looking into a crystal ball of radius $\varepsilon$ over a very long time can tell you everything about the outside world.

There is a peculiar phenomenon associated with the last example. A solution of (12) existing in the cylinder (17) need not have any extension beyond the cylinder. An example is furnished by the t-independent solutions that coincide with a harmonic function of x,y that has the circle $x^2 + y^2 = a^2$ as natural boundary. If, however, an extension exists, can we conclude anything about its regularity from the regularity of u inside the cylinder (17)? An example will show that this is not the case. It can happen that a $C^2$-solution u exists in the whole xyt-space, which is real analytic in (17), without belonging even to $C^3$ in the whole space. For the example (see [12]) we write (12) in cylindrical coordinates $r, \theta, t$ as

$$u_{tt} = u_{rr} + r^{-1}u_r + r^{-2}u_{\theta\theta}.$$

We can here separate variables in an unconventional way and see that there exists for every integer n the particular entire analytic solutions

$$u_n = e^{in(t+\theta)} \; J_n(nr)$$

where $J_n$ is the Bessel function. Now $J_n(nr)$ for large n and for complex r near a real r between -1 and 1 has estimates of the form

$$|J_n(nr)| < q^n; \quad |q| < 1;$$

on the other hand

$$J_n(nr) \approx n^{-1/2} \text{ for real } r > 1.$$

If we now form the function

$$u = \sum_{n=0}^{\infty} 2^{-kn} u_{2^n} \text{ with } k = m + \frac{3}{5}$$

where m is an arbitrary positive integer, we find that

$u \in C^a$ for $r < 1$
$u \in C^m$, $u \notin C^{m+1}$ for $r = 1$
$u \in C^{m+1}$, $u \notin C^{m+2}$ for $r > 1$.

One is accustomed to seeing characteristic surfaces as loci of jumps in the derivatives of a solution. In our example, the cylindrical surface $r = 1$, across which the change in regularity occurs, is not characteristic. This is related to the fact that generally singularities propagate along bicharacteristics. (See [11], [13], [15].) The low degree of regularity of u for $r > 1$ reflects the low degree of continuous dependence in the process of continuing a solution from the cylinder (17) into the whole space and the huge errors to be expected in reconstructing the world from a crystal ball.

There exist different geometric situations where continuation preserves regularity for solutions of (12). If, for example, a region R in xyt-space is bounded by a time-like surface S together with a portion of the coordinate plane $x = 0$, then Cauchy data of class $C^\infty$ on S assure that $u \in C^\infty$ in R provided a solution u exists in R. (See [12], [14].)

## References


[1] Lipman, Bers, Theory of pseudo-analytic functions. Lecture Notes, 1953, Institute for Mathematics and Mechanics, New York University.

[2] I. N. Vekua, Generalized analytic functions. (Tr. Pergamon Press, 1962).

[3] R. Fueter, Funktionentheorie im Hyperkomplexen. Universität Zürich, 1948/49.

[4] F. John, Plane waves and spherical means applied to partial differential equations, 1955.

[5] J. Chazarain and A. Piriou, Introduction a la theorie des equations aux derivees partielles lineaires. Gauthier-Villars, Paris, 1981.

[6] F.Treves, Introduction to Pseudodifferential and Fourier Integral Operators. Plenum Press, New York, 1980.

[7] F. John, General properties of solutions of linear elliptic partial differential equations; Proc. of the Symposium on Spectral Theory and Differential Problems, Stillwater, Oklahoma, 1951, pp 113-175.

[8] L. Hörmander, Linear Partial Differential Operators. Springer Verlag, 1963.

[9] Michael Balch, A Laurent expansion for solutions of linear elliptic differential equations, Comm. Pure Appl. Math. 19, 1966, 343-352.

[10] F. John, Partial differential equations, 4th Ed. Springer Verlag, 1982.

[11] L. Hörmander, Uniqueness theorems and wave front sets for solutions of linear differential equations with analytic coefficients. Comm. Pure Appl. Math. 24, 1971, 671-704.

[12] F. John, Continuous dependence on data for solutions of partial differential equations with analytic coefficients. Comm. Pure Appl. Math. 24, 1971, 671-704.

[13] J. Boman, On the propagation of analyticity of solutions of differential equations with constant coefficients, Ark. für Mathematik 5, 1964, 271-279.

[14] L. Hörmander, On the singularities of solutions of partial differential equations with constant coefficients. Israel J. Math. 13, 1972, 82-105.

[15] L. Hörmander, The analysis of linear partial differential operators, I. Springer Verlag, 1983.

# Remarks on Zero Viscosity Limit for Nonstationary Navier-Stokes Flows with Boundary

By Tosio Kato

Math Department
University of California
Berkeley, California

## 1. Introduction

This paper is concerned with the question of convergence of the nonstationary, incompressible Navier-Stokes flow $u = u_\nu$ to the Euler flow $\bar{u}$ as the viscosity $\nu$ tends to zero. If the underlying space domain is all of $R^m$, the convergence has been proved by several authors under appropriate assumptions on the convergence of the data (initial condition and external force); see Golovkin [1] and McGrath [2] for $m = 2$ and all time, and Swann [3] and the author [4,5] for $m = 3$ and short time. The case $m \geqslant 4$ can be handled in the same way; in fact, the simple method given in [5] applies to any dimension. All these results refer to strong solutions (or even classical solutions, depending on the data) of the Navier-Stokes equation.

The problem becomes extremely difficult if the space domain $\Omega \subset R^m$ has nonempty boundary $\partial\Omega$, due to the appearance of the boundary layer, and remains open as far as the author is aware. Here it is necessary, in general, to consider weak solutions u of the Navier-Stokes equation, since strong solutions are known to exist only for a short time interval that tends to zero as $\nu \to 0$ (except for $m = 2$), while weak solutions are known to exist for all time for any initial data in $L^2(\Omega)$, although their uniqueness is not known.

The purpose of this paper is to give some necessary and sufficient conditions for the convergence to take place. In particular, we shall show that, roughly speaking, $u \to \bar{u}$ in $L^2(\Omega)$, uniformly in $t \in [0,T]$, if and only if the energy dissipation for u during the interval [0,T] tends to zero. Here [0,T] is an interval on which the smooth solution $\bar{u}$ of the Euler equation exists, and u is any weak solution of the Navier-Stokes equation. Such results will give no ultimate solution to the problem, but they are hoped to be useful for further investigation of the problem.

## 2. Statement of Theorems

In what follows $\Omega$ is a bounded domain in $R^m$ with smooth boundary $\partial\Omega$. The Navier-Stokes equation for an incompressible fluid with density one may be written formally

$$\text{(NS)} \quad \begin{aligned} &\partial_t u - \nu\Delta u + (u\cdot\text{grad})u + \text{grad } p = f, \\ &\text{div } u = 0, \ u_{|\partial\Omega} = 0, \end{aligned}$$

where $u = u(t,x)$ is the velocity field, p the pressure, f the external force, $\nu > 0$ the (kinematic) viscosity, and $\partial_t = \partial/\partial t$. We assume ($f = f_\nu$ may depend on $\nu$)

$$(2.1) \quad f \in L^1((0,T);L^2(\Omega)) \text{ for any } T > 0.$$

Here and in what follows $L^2(\Omega)$ may denote, indiscriminately, the $L^2$-space of scalar, vector, or tensor-valued functions and similarly for other function spaces such as $C^k(\Omega)$, $H^s(\Omega)$ (Sobolev spaces).

A <u>weak solution</u> u to (NS) is assumed to satisfy the following conditions, where V is the space of vector-valued $H_0^1(\Omega)$-functions with divergence zero. (We write $\partial_k = \partial/\partial x_k$, and ( , ) [$\|\ \|$] denotes the (formal) inner product [norm] in $L^2(\Omega)$.)

$$(2.2) \quad u \in C_w([0,T];L^2(\Omega)) \cap L^2((0,T);V) \text{ for any } T > 0.$$

$$(2.3) \quad \|u(t)\|^2/2 + \nu\int_0^t \|\text{grad } u\|^2 dt \leq \|u(0)\|^2/2 + \int_0^t (f,u)dt.$$

$$(2.4) \quad \begin{aligned} (u(t),\emptyset(t)) - (u(0),\emptyset(0)) = \int_0^t &[(uu,\text{grad } \emptyset) - \\ &\nu(\text{grad } u,\text{grad } \emptyset) + (f,\emptyset) + (u,\partial_t\emptyset)]\,dt \end{aligned}$$

for every vector-valued test function $\emptyset \in C^1([0,\infty] \times \bar{\Omega})$ satisfying div $\emptyset = 0$ and vanishing on $\partial\Omega$.

Remark 2.1

(a) It is known (see Leray [6], Hopf [7], Ladyzenskaya [8], Lions [9], Temam [10]) that a weak solution u exists for any $u(0) \in L^2(\Omega)$ with div $u(0) = 0$. We assume that for each $\nu > 0$, one such weak solution $u = u_\nu$ of (NS) has been chosen. For simplicity the parameter $\nu$ and the space variable x are suppressed, as is the time variable t frequently (as in the integrand in (2.4)).

(b) $C_w$ in (2.2) indicates weak continuity.

(c) In (2.4) the following short-hand notation is used:

$$(2.5) \qquad (uu, \text{grad } \emptyset) = \sum_{j,k} (u_j u_k, \partial_k \emptyset_j) = -\sum_k (u_k \partial_k u_j, \emptyset_j).$$

(d) The test function $\emptyset$ in (2.4) is sometimes (as in Hopf [7]) assumed to have spatial compact support. To admit more general $\emptyset$ stated above, we may use the fact that (spatial) test functions with compact supports are dense in $W_0^{1,p}(\Omega)$ with divergence zero, for any $p < \infty$, which can be proved by the "pulling-in" method given by Heywood [11]. Indeed, the functions uu, grad u, etc., appearing in (2.4) belong (for each fixed t) to some $L^q(\Omega)$ with $q > 1$.

(e) (2.3) describes the energy inequality. Note that we do not require that the energy inequality hold on intervals $[t_0, t]$ with $t_0 > 0$, a condition necessary in other problems related to (NS) such as Leray's structure theorem for turbulent solutions.

The Euler equation is obtained from (NS) by formally setting $\nu = 0$. In general ($m \geqslant 3$), only local (in time) solutions are known for the Euler equation. We denote by $\bar{u}$ such a solution:

$$\text{(E)} \qquad \partial_t \bar{u} + (\bar{u} \cdot \text{grad})\bar{u} + \text{grad } \bar{p} = \bar{f}, \; 0 \leqslant t \leqslant \bar{T} < \infty.$$
$$\text{div } \bar{u} = 0, \; \bar{u}_{n \,|\, \partial\Omega} = 0 \text{ (normal component).}$$

Existence and uniqueness of a smooth local solution $\bar{u}$ have been proved by many authors (Ebin-Marsden [12], BourguignonBrezis [13], Temam [14], Lai [15], Kato [16], Kato-Lai [17], and others). Thus, we may assume

(2.6) $\bar{u}, \bar{p}, \bar{f} \in C^1([0,\bar{T}] \times \bar{\Omega})$.

We note that $\bar{T}$ may be taken arbitrarily large if $m = 2$ and $\bar{u}(0) \in C^{1+\varepsilon}(\bar{\Omega})$, div $\bar{u}(0) = 0$.

We are now able to state the main theorem.

*Theorem I*

*Fix* $T > 0$, $T \leqslant \bar{T}$, *and assume*

(2.7) $u(0) \to \bar{u}(0)$ *in* $L^2(\Omega)$ *as* $\nu \to 0$.

(2.8) $\int_0^T \|f - \bar{f}\| dt \to 0$ *as* $\nu \to 0$.

*Then the following conditions* (i) *to* (iv) *are equivalent. (All limiting relations refer to* $\nu \to 0$.)

(i) $u(t) \to \bar{u}(t)$ *in* $L^2(\Omega)$, *uniformly in* $t \in [0,T]$

(ii) $u(t) \to \bar{u}(t)$ *weakly in* $L^2(\Omega)$ *for each* $t \in [0,T]$.

(iii) $\nu \int_0^T \|\text{grad } u\|^2 dt \to 0$.

(iii') $\nu \int_0^T \|\text{grad } u\|^2_{\Gamma_{c\nu}} dt \to 0$,

*where* $\| \ \|_{\Gamma_{c\nu}}$ *denotes the* $L^2(\Gamma_{c\nu})$*-norm,* $\Gamma_{c\nu} \subset \Omega$ *being the boundary strip of width* $c\nu$, *with* $c > 0$ *fixed but arbitrary.*

*If in particular* $\bar{f} = 0$, *these conditions are equivalent to*

(iv) $u(T) \to \bar{u}(T)$ *weakly in* $L^2(\Omega)$.

*Theorem Ia*

*Replace* $L^1$ *by* $L^2$ *in* (2.1), *and replace* (2.8) *by*

(2.8a) $\int_0^{T'} \|f-\bar{f}\|^2 dt \to 0$ *for some* $T' > T$, $T' \leqslant \bar{T}$.

*Then the equivalent conditions* (i) *to* (iii) *in Theorem I are implied by*

(v) $\int_0^{T'} \|u - \bar{u}\|^2 dt \to 0.$

Remark 2.2

(a) Condition (iii) states that the energy dissipation during a finite time tends to zero as $\nu \to 0$, and (iii') states that the dissipation within a boundary strip of width $c\nu$ tends to zero.

(b) From the practical point of view, these conditions do not appear very helpful in deciding whether or not the convergence (i) takes place. In fact, we do not know whether (iii) or (iii') is always true, always false (except in trivial cases), or both possibilities exist. It may be noted, however, that if the convergence does not take place, the energy dissipation within the boundary layer of width $c\nu$ must remain finite as $\nu \to 0$. Since the boundary layer is believed to have thickness proportional to $(\nu t)^{1/2}$, this suggests that something violent must have happened for small $t > 0$.

(c) In Theorems I and Ia, the family $\{u\}$ with the continuous parameter $\nu$ may be replaced by a sequence $\{u^n\}$ corresponding to a sequence $\nu_n \to 0$ of the parameter $\nu = \nu_n$.

## 3. Proof of Easier Parts of the Theorems

First we deduce simple consequences of the properties of the solutions $u$ and $\bar{u}$. For simplicity we use the notation $|||\ \ |||_p$ for the $L^p((0,T);L^2(\Omega))$-norm, and $((\ ,\ ))$ for the (formal) scalar product in $L^2((0,T);L^2(\Omega))$. These are used indiscriminately for scalar, vector, and tensor-valued functions. K denotes various constants independent of $\nu$.

It is well known (and is easy to prove) that (2.3), (2.7) and (2.8) imply

(3.1) $|||u|||_\infty \leqslant \|u(0)\| + |||f|||_1 \leqslant K,$

(3.2) $\nu\ |||\text{grad } u|||_2^2 \leqslant \|u(0)\|^2/2 + |||u|||_\infty |||f|||_1 \leqslant K.$

Similarly, (E) and (2.6) imply

$$(3.3) \quad (\bar{u}\bar{u}, \text{grad } \bar{u}) = 0,$$

$$(3.4) \quad \|\bar{u}\|^2/2 = \|\bar{u}(0)\|^2/2 + \int_0^t (\bar{f},\bar{u})dt.$$

Now we shall prove simple implications contained in the theorems.

(a) (i) implies (ii). This is trivial.

(b) (ii) implies (iii). If (ii) is true, (2.3) gives

$$(3.5) \quad \limsup \nu |||\text{grad } u|||_2^2$$
$$\leqslant \limsup [((f,u)) - (\|u(T)\|^2 - \|u(0)\|^2/2]$$
$$\leqslant ((\bar{f},\bar{u})) - (\|\bar{u}(T)\|^2 - \|\bar{u}(0)\|^2)/2 = 0$$

by (2.7), (2.8), (3.1) and (3.4); note that $\liminf \|u(T)\| \geqslant \|\bar{u}(T)\|$ because $u(T) \to \bar{u}(T)$ weakly. To see that $((f,u)) \to ((\bar{f},\bar{u}))$, use dominated convergence in t.

(c) (iii) implies (iii') trivially.

(d) (iv) implies (iii) if $\bar{f} = 0$, since only $u(T) \to \bar{u}(T)$ weakly is needed in the above proof in (b). Indeed, $|||f|||_1 \to 0$ implies $((f,u)) \to 0$ by (3.1).

(e) (v) implies (iii) under the additional assumptions stated. To see this, we integrate (2.3) in $t \in (0,T')$ to obtain

$$(3.6) \quad \limsup \nu \int_0^{T'} (T'-t)\|\text{grad } u\|^2 dt$$
$$\leqslant \limsup \left[\int_0^{T'} (T'-t)(f,u)dt + (T'\|u(0)\|^2 - |||u|||_2^2)/2\right],$$

where $||| \; |||_2$ is taken on $(0,T')$. If (2.8a) and (v) are true, then $\int (T'-t)[(f,u)-(\bar{f},\bar{u})]dt \to 0$ and $|||u|||_2 \to |||\bar{u}|||_2$ so that the right member of (3.6) does not exceed

$$\int_0^{T'} (T'-t)(\bar{f}-\bar{u})dt + (T'\|\bar{u}(0)\|^2 - |||\bar{u}|||_2^2)/2$$

$$= \int_0^{T'} [\int_0^t (\bar{f},\bar{u})dt + (\|u(0)\|^2 - \|u\|^2)/2]dt = 0$$

by (3.4). It follows that the left member of (3.6) is zero. Since $T' > T$, this implies (iii).

## 4. Boundary Layers

In the proof of the remaining assertion that (iii') implies (i), we need a "boundary layer" v, which is a correction term (depending on $\nu$) to be subtracted from $\bar{u}$ to satisfy the zero boundary condition and which has a thin support. It may be noted at this point that v has no direct relation with the true boundary layer belonging to u. In fact, the latter is virtually unknown in the mathematical sense.

To construct v, we first introduce a smooth "vector potential" $\bar{a}$, defined on $[0,T] \times \bar{\Omega}$, such that

(4.1) $\bar{u} = \operatorname{div} \bar{a}$ on $\partial\Omega$, $\bar{a} = 0$ on $\partial\Omega$.

$\bar{a}$ is a skew-symmetric tensor of second rank, and div $\bar{a}$ is a vector with components $\sum_k \partial_k \bar{a}_{jk}$. The existence of such an $\bar{a}$ will be proved in Appendix.

We next introduce a smooth cut-off function $\zeta : \mathbb{R}^+ \to \mathbb{R}^+$ such that

(4.2) $\zeta(0) = 1$, $\zeta(r) = 0$ for $r \geqslant 1$,

and set

(4.3) $z = z(x) = \zeta(\rho/\delta)$, where $\rho = \operatorname{dist}(x,\partial\Omega)$,

with a small parameter $\delta > 0$, which is assumed to tend to zero with $\nu$ with a rate to be determined below.

The boundary layer v is defined by

$$v = \operatorname{div}(z\bar{a}) = z \operatorname{div} \bar{a} + \bar{a} \cdot \operatorname{grad} z, \tag{4.4}$$

where $\bar{a} \cdot \operatorname{grad} z$ is a vector with components $\sum_k \bar{a}_{jk} \partial_k z$. Thus, v has a thin support near $\partial\Omega$ and satisfies

$$v = \bar{u} \text{ on } \partial\Omega, \ \operatorname{div} v = 0 \text{ in } \Omega. \tag{4.5}$$

(Note that div div b = 0 if b is skew-symmetric.)

The following estimates for v can easily be established.

$$\|v\|_{L^\infty} \leqslant K, \ \|v\| \leqslant K\delta^{1/2}, \ \|\partial_t v\| \leqslant K\delta^{1/2}, \tag{4.6}$$

$$\|\operatorname{grad} v\|_{L^\infty} \leqslant K\delta^{-1}, \ \|\operatorname{grad} v\| \leqslant K\delta^{-1/2},$$

$$\|\rho \operatorname{grad} v\|_{L^\infty} \leqslant K, \ \|\rho^2 \operatorname{grad} v\|_{L^\infty} \leqslant K\delta,$$

$$\|\rho \operatorname{grad} v\| \leqslant K\delta^{1/2}.$$

Indeed, these estimates are obviously true for v replaced with z, together with analogous estimates involving second derivatives of z. Then (4.6) follows easily because $\bar{a} = 0$ on $\partial\Omega$. $\bar{a}$ and $\partial_t \bar{a}$ are smooth and vanish on $\partial\Omega$.

## 5. Proof of (iii') ⇒ (i)

We now assume (iii') and estimate $\|u-\bar{u}\|^2$ using (2.3), (2.7) and (3.4):

$$\begin{aligned} \|u-\bar{u}\|^2 &= \|u\|^2 + \|\bar{u}\|^2 - 2(u,\bar{u}) \\ &\leqslant \|u(0)\|^2 + 2\int_0^t (f,u)dt + \|\bar{u}(0)\|^2 + 2\int_0^t (\bar{f},\bar{u})dt - 2(u,\bar{u}) \\ &\leqslant o(1) + 2\int_0^t [(f,u) + (\bar{f},\bar{u})]dt + 2\|\bar{u}(0)\|^2 - 2(u,\bar{u}-v), \end{aligned} \tag{5.1}$$

where v is defined in the previous section and where o(1) denotes a quantity that tends to zero as $\nu \to 0$ uniformly in $t \in [0,T]$. Note that $\|v\| \leqslant K\delta^{1/2}$ by (4.6) and $\delta \to 0$ with $\nu$, and that $\|u\|$

$\leqslant$ K by (3.1). We have introduced the boundary layer v into the last term of (5.1) in order to facilitate the following estimates.

To estimate the last term on the right of (5.1), we use $\emptyset = \bar{u} - v$ as a test function in (2.4); this is allowed since $\bar{u} - v$ is smooth with $\operatorname{div}(\bar{u}-v) = 0$ and vanishes on $\partial\Omega$. The result is, when multiplied with $-2$,

$$(5.2) \quad -2(u,\bar{u}-v) + 2\|u(0)\|^2 = o(1) + \int_0^t [-2(uu,\operatorname{grad}(\bar{u}-v))$$

$$+ 2\nu\,(\operatorname{grad} u,\operatorname{grad}(\bar{u}-v)) - 2(f,\bar{u}) - 2(u,\partial_t(\bar{u}-v))]\,dt;$$

note that $\|u(0)-\bar{u}(0)\| \to 0$ and $\|v(t)\| \to 0$.

The last term in the integrand in (5.2) is estimated as

$$(5.3) \quad \begin{aligned} -2(u,\partial_t(\bar{u}-v)) &= -2(u,\partial_t\bar{u}) + o(1) \\ &= o(1) + 2(u,(\bar{u}\cdot\operatorname{grad})\bar{u}) - 2(u,\bar{f}) \end{aligned}$$

(see (3.1), (4.6) and (E).) It follows from (5.1) to (5.3) that

$$(5.4) \quad \begin{aligned} \|u-\bar{u}\|^2 &\leqslant o(1) + 2\int_0^t [(f-\bar{f},u-\bar{u}) - (uu,\operatorname{grad}(\bar{u}-v)) \\ &\quad + (u,(\bar{u}\cdot\operatorname{grad})\bar{u}) + \nu\,(\operatorname{grad} u,\operatorname{grad}(\bar{u}-v))]\,dt \\ &\leqslant o(1) + 2\int_0^t [(f-\bar{f},u-\bar{u}) - ((u-\bar{u})(u-\bar{u}),\operatorname{grad}\bar{u}) \\ &\quad + (uu,\operatorname{grad} v) + \nu\,(\operatorname{grad} u,\operatorname{grad}(\bar{u}-v)]\,dt, \end{aligned}$$

where we have used the equality

$$(5.5) \quad (uu,\operatorname{grad}\bar{u}) - (u,(\bar{u}\cdot\operatorname{grad})\bar{u}) = ((u-\bar{u})(u-\bar{u}),\operatorname{grad}\bar{u}),$$

which follows easily from $\operatorname{div} u = \operatorname{div}\bar{u} = 0$, $u \in H_0^1(\Omega)$, $\bar{u}_{n\,|\,\partial\Omega} = 0$.

In view of the simple inequalities

$$(f-\bar{f},u-\bar{u}) \leqslant \|f-\bar{f}\| \ \|u-\bar{u}\| \leqslant K\|f-\bar{f}\|,$$
$$((u-\bar{u})(u-\bar{u}),\text{grad } \bar{u}) \leqslant K\|u-\bar{u}\|^2,$$

we obtain from (5.4) the following integral inequality:

$$(5.6) \qquad \|u-\bar{u}\|^2 \leqslant o(1) + \int_0^t [K\|u-\bar{u}\|^2 + R(t)]dt,$$

where

$$(5.7) \qquad R(t) = (uu,\text{grad } v) + \nu\ (\text{grad } u,\text{grad}(\bar{u}-v)) + K\|f-\bar{f}\|.$$

The integral inequality (5.6) for $\|u-\bar{u}\|^2$ is of a familiar type. It will lead to the desired result $\|u-\bar{u}\|^2 = o(1)$ if we can show that

$$(5.8) \qquad \int_0^t R(t)\ dt \leqslant o(1).$$

To prove (5.8), we first note that

$$|(uu,\text{grad } v)| \leqslant \|\rho^{-1}u\|^2_{\Gamma_\delta} \|\rho^2 \text{grad } v\|_{L^\infty} \leqslant K\delta\|\text{grad } u\|^2_{\Gamma_\delta}$$

by (4.6) and the well-known inequality of Hardy-Littlewood (note that $u \in H_0^1(\Omega)$); we can take $\|\rho^{-1}u\|_{\Gamma_\delta}$ only on the boundary strip $\Gamma_\delta$ because v is supported on $\Gamma_\delta$. Similarly,

$$|\nu(\text{grad } u,\text{grad}(\bar{u}-v))|$$
$$\leqslant \nu\ \|\text{grad } u\|\ \|\text{grad } \bar{u}\| + \nu\ \|\text{grad } u\|_{\Gamma_\delta}\ \|\text{grad } v\|_{\Gamma_\delta}$$
$$\leqslant K\nu\ \|\text{grad } u\| + K\nu\delta^{-1/2}\ \|\text{grad } u\|_{\Gamma_\delta}$$

by (4.6).

If we simply set $\delta = c\nu$, we thus obtain

$$R(t) \leqslant K\nu \, \|\text{grad } u\|^2_{\Gamma_{c\nu}} + K\nu \, \|\text{grad } u\| + K\nu^{1/2} \, \|\text{grad } u\|_{\Gamma_{c\nu}}$$

$$+ K \, \|f-\bar{f}\|.$$

From this (5.8) follows by (iii') and (2.8), since

$$\int_0^t \nu \, \|\text{grad } u\| dt \leqslant t^{1/2} \, \nu \, |\!|\!|\text{grad } u|\!|\!| = O(\nu^{1/2})$$

by (3.2).

## Appendix
## Construction of the Vector Potential

*Lemma* A1

*Let* u *be a smooth tangential vector field on a smooth closed surface* $\Gamma$ *in* $R^m$. *There exists a skew-symmetric tensor field* a *of second rank on* $R^m$ *such that* a = 0 *and* div a = u *on* $\Gamma$ ($\partial_k a_{jk} = u_j$ *in tensor notation*). *If* u *depends on a parameter* t *smoothly,* a *can be chosen similarly.*

Proof

If $\Gamma$ is the plane $x_1 = 0$, $u = u(x')$ is defined for $x' = (x_2,\dots,x_m)$ with $u_1 = 0$. If we set $a_{j1} = -a_{1j} = x_1 u_j(x')$, $a_{11} = 0$, and $a_{jk} = 0$ for $j, k \geqslant 2$, a satisfies the required conditions.

In the general case, the problem is locally reduced to the special case just considered by a coordinate transformation with Jacobian determinant 1. Then we may identify u and a with an (m−1) form and an (m−2) form, respectively, so that div a = u has an invariant meaning da = u. Thus, we can construct an a with the required properties in a neighborhood of each point of $\Gamma$.

Next we construct a in a neighborhood of $\Gamma$. To this end we use a partition of unity $\{\varnothing^s\}$ in a neighborhood of $\Gamma$ such that on the support of each $\varnothing^s$, a local solution $a^s$ can be constructed as above. Setting $a = \sum_s \varnothing^s a^s$ then gives the desired a. Indeed, it is obvious that $a = 0$ on $\Gamma$, and $\operatorname{div} a = \sum_s [\varnothing^s \operatorname{div} a^s + (\operatorname{grad} \varnothing^s) a^s]$ $= \sum_s \varnothing^s u = u$ on $\Gamma$.

Finally, we extend $a = a_0$ thus obtained to all of $R^m$. It suffices to introduce a smooth cut-off function $\zeta$ and set $a = \zeta a_0$. Here $\zeta$ should be equal to 1 in a neighborhood of $\Gamma$ and have support contained in the domain of $a_0$.

*Corollary*

u *can be extended to a vector field on* $R^m$ *with* $\operatorname{div} u = 0$ *and* $u_{n|\Gamma} = 0$.

<u>Proof</u>

It suffices to set $u = \operatorname{div} a$ (note that $\operatorname{div} \operatorname{div} a = 0$).

## References


1. K. K. Golovkin, Vanishing viscosity in Cauchy's problem for hydronamics equations, Proc. Steklov Inst. Math. (English translation) 92 (1966), 33–53.
2. F. J. McGrath, Non-stationary plane flow of viscous and ideal fluids, Arch. Rational Mech. Anal. 27 (1968), 329–348.
3. H. Swann, The convergence with vanishing viscosity of non-stationary Navier-Stokes flow to ideal flow in $R_3$, Trans. Amer. Math. Soc. 157 (1971), 373–397.
4. T. Kato, Non-stationary flows of viscous and ideal fluids in $R^3$, J. Functional Anal. 9 (1972), 296–305.
5. T. Kato, Quasi-linear equations of evolution, with applications to partial differential equations, Lecture Notes In Math. 448, Springer 1975, 25–70.
6. J. Leray, Sur le mouvement d'un liquide visqueux emplissant l'espace, Acta Math. 63 (1934), 193–248.
7. E. Hopf, Über die Anfangswertaufgabe für die hydrodynamischen Grundgleichungen, Math. Nachr. 4 (1951), 213–231.
8. O. A. Ladyzenskaya, The mathematical theory of viscous incompressible flow, Second English Edition, Gordon and Breach, New York 1969.

9. J. L. Lions, Quelque méthodes de résolution des problèmes aux limites non linéaires, Dunod, Gauthier-Villars, Paris 1969.

10. R. Temam, Navier-Stokes equations, North-Holland, Amsterdam-New York-Oxford 1979.

11. J. Heywood, On uniqueness questions in the theory of viscous flow, Acta Math. 136 (1976), 61-102.

12. D. G. Ebin and J. Marsden, Groups of diffeomorphisms and the motion of an incompressible fluid, Ann. Math. 92 (1970), 102-163.

13. J. P. Bourguignon and H. Brezis, Remarks on the Euler equation, J. Functional Anal. 15 (1974), 341-363.

14. R. Temam, On the Euler equations of incompressible perfect fluids, J. Functional Anal. 20 (1975), 32-43.

15. C. Y. Lai, Studies on the Euler and the Navier-Stokes equations, Thesis, University of California, Berkeley, 1975.

16. T. Kato, On classical solutions of the two-dimensional non-stationary Euler equation, Arch. Rational Mech. Anal. 25 (1967), 188-200.

17. T. Kato and C. Y. Lai, Nonlinear evolution equations and the Euler flow, J. Functional Anal. (1984), to appear.

# Free Boundary Problems in Mechanics*

By Joseph B. Keller
Departments of Mathematics
and Mechanical Engineering
Stanford University
Stanford, California 94305

Abstract

Free boundary problems are defined and illustrated by several problems in mechanics. First the problem of finding the free surface of a liquid in hydrostatic equilibrium is considered. Then the effect of surface tension is taken into account. Finally, the contact of an inflated membrane, such as a balloon or tire, with a solid surface is formulated. This problem is solved by the method of matched asymptotic expansions when the contact area is small.

---

*Research supported by the Office of Naval Research, the Air Force Office of Scientific Research, the Army Research Office, and the National Science Foundation.

A. Introduction

Free boundary problems are a special class of boundary value problems for differential equations in a domain. In them all or part of the boundary of the domain must be found along with the solution in the interior of the domain. Such problems are very common in continum mechanics; i.e., the mechanics of fluids and solids. They occur whenever there is an interface between a moving or stationary liquid or solid and the adjacent medium, since the location of the interface is generally unknown before the problem is solved. Problems involving the shape of liquid drops and liquid jets, the motion of waves on the surface of a liquid, and the motion of a bubble in a liquid are examples of free boundary problems in hydrostatics and hydrodynamics. The determination of the deformed shape of a stationary or moving solid body is an example in solid mechanics.

The first step in considering these or any other physical problems is their formulation based upon the relevant physical laws. The second step is to determine whether the formulation is wrong. This will be the case if no solution exists when experience suggests that one should exist, if more than one solution exists when experience suggests that there should be just one, or if the solution does not depend continuously on the data of the problem when experience suggests that it should depend continuously. If it passes all three tests we say the problem is properly formulated or well posed. This does not mean that it is the correct formulation of the physical problem but only that it is not obviously incorrect. The only way to tell whether it is correct is to compare the solution of the problem with experimental results. Therefore, the third and main step is to find or calculate the solution and examine its properties.

Many mathematicians who consider physical problems stop after the second step, that of proving that the problems are well posed. This is unfortunate because it is only after that that the problems become interesting from a scientific or technical point of view.

In this lecture I will formulate a number of different free boundary problems in mechanics and show how to solve them.

## B. Hydrostatics

### 1. Formulation

Hydrostatics is the study of the mechanics of a liquid at rest under the influence of external forces. The basic fact underlying the subject is that the force per unit area exerted by the fluid lying on one side of a surface upon that lying on the other side is normal to the surface, and its magnitude is independent of the direction of the normal. This magnitude is called the pressure $p(x)$, which may be a function of position $x$. It follows that the equation of force balance, which must hold in order that the fluid remain at rest, is

$$(1) \qquad \nabla p(x) = f(x), \qquad x \in D$$

Here $f(x)$ is the external force per unit volume, which is assumed to be known, and $D$ is the domain containing the liquid.

A typical problem is that of finding the configuration of a given volume $V$ of liquid in a container with surface $S$, and of finding the pressure in the liquid. We shall assume that the container is open to the air and that the pressure in the air is a constant $p_0$. Then the liquid will be bounded in part by the surface $S$ and in part by some unknown surface $S'$ at which the liquid is in contact with the air. This surface $S'$ is called a free boundary of the domain $D$ containing the liquid. If surface tension is negligible, the pressure must be continuous across $S'$, so we have

$$(2) \qquad p(x) = p_0, \qquad x \in S'$$

In addition, the fact that the liquid has volume $V$ yields

$$(3) \qquad \int_D dx = V$$

Thus, the problem is to find $S'$ and $p(x)$ such that $S'$ and part of $S$ together bound a domain $D$ of volume $V$ within which $p(x)$ satisfies (1), and on $S'$, $p(x)$ satisfies (2).

## 2. Equilibrium in a Uniform Gravitational Field

Let us take the external force to be the force of gravity $f(x) = -(0,0,\rho g)$ which acts parallel to the $-z$ axis. Its magnitude is the product of the liquid density $\rho$ and the acceleration of gravity $g$. Then the $x$ and $y$ components of (1) show that $p$ is independent of $x$ and $y$, while the $z$-component becomes $p_z = -\rho g$. Integration yields

$$(4) \qquad p = -\rho g z + c$$

Here $c$ is a constant of integration to be determined.

We now use (4) in (2) and find that

$$(5) \qquad z = (p_0 - c)/\rho g, \qquad x \in S'$$

This equation (5) shows that $S'$ is a portion of a horizontal plane. Finally, we use this fact in (3) to determine the height of the plane, and thus we find $c$.

If the container is too small it may not bound a domain of volume $V$ with any horizontal plane, in which case the problem will have no solution. This is the case when we try to pour too much liquid into a small glass. Alternatively, there may be many solutions when the domain $D$ is not connected. This occurs if "the container" is really two containers, such as two glasses. Then (4) will hold with separate constants, $c_1$ and $c_2$, in each container, and the liquid can be divided between them in infinitely many ways.

## 3. Surface Tension

The interface between a liquid and the air exhibits the property of surface tension. This property results in a pressure drop across a curved interface, proportional to the mean curvature $K_m$ of the interface. Thus, when this jump is taken into account, (2) must be replaced by

$$(2') \qquad p(x) = p_0 + 2TK_m(x), \qquad x \in S'$$

Here the constant T is the coefficient of surface tension, which is characteristic of the two fluids, liquid and air, in contact across the interface.

In addition, the interface S' and the container wall S must meet at a fixed angle $\beta$ characteristic of the wall material, the liquid and the gas. We can write this condition as

$$(6) \qquad n'(x) \cdot n(x) = \cos\beta, \qquad x \in S' \cap S$$

Here n'(x) is the unit normal to S' and n(x) is the unit normal to S at any point x on the curve of intersection of S' and S. Now the hydrostatic problem is to find S' and p(x) such that (1), (2'), (3) and (6) hold.

In the absence of external forces f(x) = 0 in (1), which implies that p(x) is a constant $p_1$. Then (2') shows that

$$(7) \qquad K_m(x) = (p_1 - p_0)/2T, \qquad x \in S'$$

Thus S' is a surface of constant mean curvature. The constant pressure $p_1$, which determines the magnitude of the curvature, is to be determined so that the volume condition (3) is satisfied.

As an example, suppose the liquid does not touch the container, which is possible in the absence of external forces. Then S' is a closed surface, so it must be a sphere of radius

$$(8) \qquad R = 2T/(p_1 - p_0)$$

The radius R, and thus the pressure $p_1$, are determined by (3). This solution is not unique because the center of the sphere can be anywhere, provided the sphere does not touch the container. If there is no such location within the container a different solution, in which the liquid touches the container, must be sought.

## C. Contact of an Inflated Membrane With a Rigid Surface

### 1. Introduction

Consider a closed inflated membrane, such as a rubber balloon or an inner tube. When pushed against a rigid surface, the membrane deforms and some part of it comes into contact with the surface. The shape of the deformed membrane cannot be found without simultaneously finding the curve $\Gamma$ which bounds the contact region. Thus, this boundary curve is a free boundary.

The shape of the deformed membrane is governed by the equation of mechanical equilibrium. This equation must be supplemented by some condition describing the way in which the membrane is pushed toward the rigid surface. Along $\Gamma$ the membrane must be in contact with the rigid surface. These conditions would suffice to determine the membrane shape if $\Gamma$ were given. However, mechanical equilibrium requires also that the membrane be tangent to the rigid surface along $\Gamma$. With this additional condition, the preceding equation and conditions determine $\Gamma$ as well as the membrane shape.

### 2. Geometrical Formulation

To illustrate the formulation of such a problem, we shall consider the rather artificial case of a membrane attached to the rigid plane $z = H$ along a closed curve $\tilde{C}$. We suppose that the space between the membrane and the plane is inflated to a pressure P above the pressure of the surrounding air. As a consequence of this inflation, the membrane will be stretched until a certain constant tension $\sigma$ develops in it. The value of $\sigma$ depends upon the thickness, the elastic properties and the amount of stretching of the membrane material. Then the normal force $2\sigma K_m$ due to the tension and the mean curvature $K_m$ of the membrane will just balance the pressure difference P across the membrane.

This balance yields the equation of mechanical equilibrium

$$(1) \qquad 2\sigma K_m = P$$

We must supplement (1) by the requirement that the edge of the membrane lie on $\tilde{C}$. Thus, the membrane is a surface with the constant mean curvature $P/2\sigma$, bounded by the curve $\tilde{C}$.

Let us suppose that another rigid plane $z = 0$ is tangent to the membrane at its lowest point, which we choose as the origin 0 of cartesian coordinates. Now we push the upper plane downward a distance $h < H$, so that the edge $\tilde{C}$ lies in the plane $z = H - h > 0$. We assume that the membrane will come into contact with the bottom plane $z = 0$ within a region around the origin bounded by a curve $\Gamma$. This curve is to be found along with the shape of the deformed membrane. To find them we must find a surface satisfying (1) bounded by $\Gamma$ and $\tilde{C}$, and tangent to the plane $z = 0$ along $\Gamma$.

3. Analytical Formulation

We shall now assume that the deformed surface can be represented by the equation $z = u(x,y)$. Here x,y lies in the annular region D between $\Gamma$ and C, the projection of $\tilde{C}$ into the plane $z = o$. In terms of u, (1) becomes

(2) $$\nabla\circ([1 + (\nabla u)^2]^{-1/2}\nabla u) = P/\sigma \qquad \text{in } D$$

The boundary conditions are

(3) $$u = H - h \text{ on } C$$

(4) $$u = 0 \text{ on } \Gamma$$

(5) $$\partial u/\partial n = 0 \text{ on } \Gamma$$

Equations (2)–(5) constitute the problem for the determination of u and $\Gamma$.

4. Axially Symmetric Case

As an example, let us consider the case in which the curve $\tilde{C}$ is a circle of radius R so that C is also a circle of radius R. Then before the upper plane is pushed down, i.e., when $h = 0$, the undeformed surface is a portion of a sphere of radius $\rho$. From (2) we find that $\rho = P/2\sigma$. In order that the surface pass through $\tilde{C}$, its radius must be less than $\rho$, so $R < \rho = 2\sigma/P$. Then the lowest point of the surface lies a distance $H = \rho - (\rho^2 - R^2)^{1/2}$ below the curve $\tilde{C}$. The region of contact between the surface and the plane $z = 0$ is just a single point, the origin, so we may say that $\Gamma$ is the origin when $h = 0$.

For $h > 0$ we expect that $\Gamma$ and the surface both will be symmetric about the z-axis. This symmetry follows from the symmetry of the problem if it has a unique solution. Assuming symmetry, we have $u = u(r)$, where r is the radial coordinate in the x,y plane, and $\Gamma$ is a circle, the radius of which we call a. Then (2)–(5) become, with $P/\sigma$ replaced by $2\rho^{-1}$,

(6) $$r^{-1}\partial_r(r[1 + u_r^2]^{-1/2}u_r) = 2\rho^{-1}$$

(7) $$u(R) = H - h$$

(8) $$u(a) = 0$$

(9) $$u_r(a) = 0$$

To solve (6) we multiply both sides by r and integrate from $r = a$ using (9) to evaluate the integration constant. This yields

(10) $$r[1 + u_r^2]^{-1/2}u_r = \rho^{-1}(r^2 - a^2)$$

Now we divide by r, square both sides, solve for $u_r^2$, take the square root and integrate from $r = a$ using (8) to get

(11) $$u(r) = \int_a^r \rho^{-1}(r^2 - a^2)\ [r^2 - \rho^{-2}(r^2 - a^2)^2]^{-1/2}\ dr$$

This axially symmetric surface of constant mean curvature was found by Delaunay in 1841.

Finally we use (11) in (7) to obtain

$$(12) \qquad H - h = \rho^{-1} \int_a^R (r^2 - a^2)\,[r^2 - \rho^{-2}(r^2 - a^2)^2]^{-1/2}\, dr$$

Now (12) can be viewed as an equation for the radius a of the free boundary $\Gamma$. Once it is solved, the surface height u(r) is given by the integral (12), which is expressible in terms of standard elliptic integrals.

5. Small Slope Approximation

When $\rho^{-1}$ is small (compared to $R^{-1}$), we can expand the integrals in (11) and (12) in powers of $\rho^{-1}$. Then we obtain

$$(13) \qquad u(r) = \rho^{-1} \int_a^r (r^2 - a^2) r^{-1}\, dr + O(\rho^{-3})$$

$$= \rho^{-1} \left[\frac{1}{2}(r^2 - a^2) - a^2 \log \frac{r}{a}\right] + O(\rho^{-3}),$$

$$(14) \qquad H - h = \rho^{-1} \left[\frac{1}{2}(R^2 - a^2) - a^2 \log \frac{R}{a}\right] + O(\rho^{-3}).$$

Thus (14) yields the relation between a and h, and (13) gives u(r).

The results (13) and (14) can be obtained also by making the assumption that $\rho^{-1}$ is small in (6). This suggests that the slope $|u_r|$ is small, so we neglect $u_r^2$ and obtain from (6)

$$(15) \qquad u_{rr} + r^{-1} u_r = 2\rho^{-1}$$

The solution of (15) which satisfies (8) and (9) is given by (13) with $O(\rho^{-3})$ omitted. This solution, when used in (7), yields (14) without $O(\rho^{-3})$. From the solution we can check that $u_r = O(\rho^{-1})$, which shows that this procedure is consistent.

The small slope approximation can be made in (2) even when C is not a circle. It is only necessary that $P/\sigma = 2\rho^{-1}$ be small compared to the maximum curvature of C. Then we neglect $(\nabla u)^2$ in (2) to obtain

(16) $\Delta u = 2\rho^{-1}$ in D

The other conditions (3)-(5) are unchanged.

## 6. Asymptotic Solution for Small Contact Regions

We shall now show how to solve the contact problem formulated above when the displacement h is small so that the contact region is also small. In doing so we shall employ the small slope approximation. Thus we seek u(x,y,h) and $\Gamma(h)$ satisfying (16) and (3)-(5) asymptotically for h small.

When h = 0, the surface z = u(x,y,0) is assumed to touch the plane z = 0 at one point, x = y = 0. Then $u_x(0,0,0) = 0$ and $u_y(0,0,0) = 0$, and we can orient the x and y axes so that

(17) $u(x,y,0) = \alpha x^2 + \beta y^2 + O(r^3)$

For h > 0, we shall represent u(x,y,h) by two different asymptotic expansions in h: an inner expansion valid near the contact region and an outer expansion valid far from this region.

Far from the contact region the effect of the contact is like that of a localized force of some unknown strength. Therefore, in the outer region we shall write u in the form

(18) $u(x,y,h) = u(x,y,0) - h + K[\log r + G(x,y)] + O(h^2)$

To find the constant K and the function G(x,y) we substitute (18) into (16) and (3), and obtain

(19) $\Delta G(x,y) = 0$ inside C

(20) $G(x,y) = -\log r$ on C

At the origin (19) does not hold because log r is singular there. We assume that G is regular everywhere, including the origin. Then we see that G is the regular part of the Green's function for Laplace's equation in the domain bounded by C. We shall assume that the curve

C is symmetric in the x and y axes, and therefore so is $G(x,y)$. Then near the origin it has the form

$$G(x,y) = G_0 + G_1(x^2 - y^2) + O(r^4) \tag{21}$$

Near the contact region we shall express u in terms of the elliptic coordinates $\mu$ and $\theta$, which are related to x and y by

$$x = \frac{a}{2} \cosh \mu \cos \theta, \qquad y = \frac{a}{2} \sinh \mu \sin \theta \tag{22}$$

Here a is the interfocal distance of any ellipse $\mu$ = constant. Now we express $u(x,y,h)$ as $u(x,y,o)$ given by (17) plus a solution of Laplace's equation, and we write it in terms of $\mu$ and $\theta$. The $x^2$ and $y^2$ terms in (17) involve the functions $\sin^2 \theta$ and $\cos^2 \theta$, which are expressible in terms of constants and $\cos 2\theta$. Therefore, we shall choose the solutions of Laplace's equation which are independent of $\theta$ or proportional to $\cos 2\theta$, since they will suffice to satisfy the boundary conditions (4) and (5).

On the basis of these considerations, we write u in the region near $\Gamma$ as

$$\begin{aligned} u(x,y,h) &= \frac{\alpha a^2}{4} \cosh^2 \mu \cos^2 \theta + \frac{\beta a^2}{4} \sinh^2 \mu \sin^2 \theta \\ &\quad - D - E\mu - (Ae^{2\mu} + Be^{-2\mu}) \cos 2\theta + \ldots \\ &= \frac{\alpha a^2}{8} \cosh^2 \mu + \frac{\beta a^2}{8} \sinh^2 \mu - D - E\mu + [\frac{\alpha a^2}{8} \cosh^2 \mu \\ &\quad - \frac{\beta a^2}{8} \sinh^2 \mu - Ae^{2\mu} - Be^{-2\mu}] \cos 2\theta + \ldots \end{aligned} \tag{23}$$

The second form of (23) follows from the first form by expanding $\sin^2 \theta$ and $\cos^2 \theta$, and rearranging terms. To impose the boundary conditions (4) and (5) we assume that $\Gamma(h)$ is nearly an ellipse, so we write its equation as $\mu = \mu_0 + O(h)$. Then (5) becomes

$$\partial u/\partial \mu + O(h) = 0 \text{ on } \mu = \mu_0 + O(h) \tag{24}$$

To satisfy (4) we set $\mu = \mu_0 + O(h)$ in the last expression in (23), and equate the result to zero. Then the terms independent of $\theta$ yield

$$(25) \qquad \frac{\alpha a^2}{8} \cosh^2 \mu_0 + \frac{\beta a^2}{8} \sinh^2 \mu_0 - D - E\mu_0 + \dots = 0$$

The terms in cos $2\theta$ yield

$$(26) \qquad \frac{\alpha a^2}{8} \cosh^2 \mu_0 - \frac{\beta a^2}{8} \sinh^2 \mu_0 - Ae^{2\mu_0} - Be^{-2\mu_0} + \dots = 0$$

Next we take the $\mu$ derivative of the last expression in (23), set $\mu = \mu_0 + O(h)$ and use the result in (24). Then the terms independent of $\theta$, and those in cos $2\theta$ yields the following equations, respectively:

$$(27) \qquad (\alpha + \beta) \frac{a^2}{4} \cosh \mu_0 \sinh \mu_0 - E + \dots = 0$$

$$(28) \qquad (\alpha - \beta) \frac{a^2}{4} \cosh \mu_0 \sinh \mu_0 - 2\, Ae^{2\mu_0} + 2\, Be^{-2\mu_0} + \dots = 0$$

Equations (25)-(28) are four equations involving the six constants a, $\mu_0$, A, B, D and E. Additional equations will result from the requirement that the inner expansion (2) must agree with the outer expansion (18) in some intermediate zone, called the overlap region. In this region x and y must be small while $\mu$ must be large. From (22) we find that for $\mu$ large

$$(29) \qquad x \sim \frac{a}{4} e^{\mu} \cos \theta, \; y \sim \frac{a}{4} e^{\mu} \sin \theta, \; r \sim \frac{a}{4} e^{\mu}, \; \mu >> 1$$

Since u(x,y,o) is common to both expansions we can subtract it from u before matching the two expansions. Thus we rewrite (18) for x and y small as follows by using (21) and then (29):

$$(30)\qquad u(x,y,h) - u(x,y,0) = -h + K \log r + KG_0 + KG_1 (x^2 - y^2) + \dots$$

$$= -h + K \log \frac{a}{4} + K\mu + KG_0$$

$$+ KG_1 \frac{a^2}{16} e^{2\mu} (\cos^2 \theta - \sin^2 \theta) + \dots$$

Next we write (23) for $\mu$ large in the form

$$(31)\qquad u(x,y,h) - u(x,y,0) = -D - E\mu - Ae^{2\mu} \cos 2\theta + \dots$$

Equating (30) and (31) leads to the following three equations from the constant terms, the terms proportional to $\mu$ and the terms proportional to $e^{2\mu} \cos 2\theta$, respectively:

$$(32)\qquad -D = -h + K \log \frac{a}{4} + KG_0$$

$$(33)\qquad -E = K$$

$$(34)\qquad -A = KG_1 \frac{a^2}{16}$$

Now (25)–(28) and (32)–(34) provide seven equations for the seven constants a, $\mu_0$, A, B, D, E and K. The constants $\alpha$, $\beta$, h, $G_0$, and $G_1$ are assumed to be known.

By adding one half times (28) to (26) we get

$$(35)\qquad A = \frac{a^2}{16} e^{-2\mu_0} [\alpha \cosh^2 \mu_0 - \beta \sinh^2 \mu_0$$

$$+ (\alpha - \beta) \cosh \mu_0 \sinh \mu_0]$$

By subtracting one half of (28) from (26) we find

$$(36) \qquad B = \frac{a^2}{16} e^{2\mu_0} [\alpha \cosh^2 \mu_0 - \beta \sinh^2 \mu_0$$

$$- (\alpha - \beta) \cosh \mu_0 \sinh \mu_0]$$

From (27) we find E and then from (33) we get K:

$$(37) \qquad -K = E = (\alpha + \beta) \frac{a^2}{4} \cosh \mu_0 \sinh \mu_0$$

Now (37) and (25) yield

$$(38) \qquad D = \frac{a^2}{8} [\alpha \cosh^2 \mu_0 + \beta \sinh^2 \mu_0$$

$$- 2\mu_0(\alpha + \beta) \cosh \mu_0 \sinh \mu_0]$$

Thus we have expressed five of the constants in terms of a and $\mu_0$.

Finally (32) and (34) lead to the following pair of equations to be solved for a and $\mu_0$:

$$(39) \qquad \frac{a^2}{8} [\alpha \cosh^2 \mu_0 + \beta \sinh^2 \mu_0$$

$$- 2\mu_0(\alpha + \beta) \cosh \mu_0 \sinh \mu_0]$$

$$= h + \frac{a^2}{4} (\alpha + \beta) \cosh \mu_0 \sinh \mu_0 [G_0 + \log \frac{a}{4}]$$

$$(40) \qquad e^{-2\mu_0} [\alpha \cosh^2 \mu_0 - \beta \sinh^2 \mu_0$$

$$+ (\alpha - \beta) \cosh \mu_0 \sinh \mu_0]$$

$$= \frac{a^2}{4} (\alpha + \beta) G_1 \cosh \mu_0 \sinh \mu_0$$

From (39) we see that $a^2$ is small, of order h. Therefore, we can neglect the right side of (40), which is proportional to $a^2$, and then (40) becomes an equation for $\mu_0$. Since the left side of (40) is also the right side of (35), we conclude that A = 0 to leading order in $a^2$.

To check these results, let us apply them to the case in which C is a circle of radius R. Then $G = G_0 = -\log R$ so that $G_1 = 0$ and $\alpha = \beta = 1/2\rho$. Now (40) becomes $\cosh^2 \mu_0 - \sinh^2 \mu_0 = 0$ which implies that $\mu_0 = \infty$. To solve (39) we let a tend to zero while $\mu_0$ tends to infinity with $ae^{\mu_0}/4 = b$ remaining constant. Then for $\mu_0$ large and a small we obtain from (39)

$$(41) \qquad \frac{b^2}{2\rho}(1 - 2\mu_0) = h + \frac{b^2}{\rho}\left(-\log R + \log \frac{a}{4}\right)$$

Upon rearrangement this becomes

$$(42) \qquad h = \frac{b^2}{\rho}\left(\log \frac{R}{b} + 1/2\right)$$

This result agrees exactly with (14) when we identify b with a, note that in (14) $H = R^2/2\rho$, and drop $O(\rho^{-3})$.

The results of this section can be derived systematically and formally by employing stretched variables to obtain the inner expansion, as is customary when applying the method of matched asymptotic expansions. We have proceeded in a more heuristic manner to avoid additional formalism and to convey the spirit behind the method.

If we are interested only in the inner expansion, we can formulate a new problem which does not involve the outer boundary C. Instead we prescribe the force F applied to the membrane. Then in the formulation we replace (7) by the condition

$$(7') \qquad F = P \cdot \text{contact area}$$

For the case when $\Gamma$ is a circle of radius a, (12) is replaced by

(12') $F = \pi a^2 P$

This equation determines a.

For the general case, we ignore the outer expansion given by (18)-(21). In the inner expansion (23) we set $A = 0$ to prevent the perturbation of the membrane displacement from growing too rapidly with distance from $\Gamma$. Then (25)-(28) and (7') are five equations for the constants a, $\mu_0$, B, D and E. From them we get (35) with $A = 0$, which determines $\mu_0$, and (36)-(38) for B, E and D in terms of $\mu_0$ and a. Finally (7') becomes the following equation for a:

(43) $$F/P = \pi \frac{a^2}{4} \cosh \mu_0 \sinh \mu_0$$

The results for the inner expansion in the previous problem are the same as the present ones, since in both cases $A = 0$, and (36)-(38) are the same. Of course (39) for a still differs from (43) for a because in the first case the displacement is prescribed while in the second case the force is prescribed. When the displacement of C is prescribed, it is necessary to find both the inner and outer expansions of the solution and to match them. Only in this way can we determine a and the other constants in the inner solution. However, as the preceding paragraph shows, when the force on C is prescribed, the constants in the inner expansion can be determined without matching to the outer expansion.

## References

Free boundary problems in fluid mechanics are discussed in many texts on hydrodynamics. An extensive presentation is given in:

H. Lamb, "Hydrodynamics," Dover, New York, 1945.

A fine monograph entirely concerned with such problems is:

G. Birkhoff and E. Zarantonello, "Jets, wakes and cavities," Academic Press, New York, 1957.

The problem of contact of two elastic bodies was solved by:

H. Hertz, Jour. für Math. (Crelle) Bd. 92, 1881.

His analysis concerns only the inner expansion. It is reproduced in various texts on elasticity, such as:

A. E. H. Love, "A treatise on the mathematical theory of elasticity," Dover, New York, 1944, pp. 193-198.

Contact of a two-dimensional membrane with a rigid surface has been treated in the following two papers:

C. H. Wu, "On the contact problems of inflated cylindrical membranes with a life raft as an example," J. Appl. Mech. 38, 615-622, 1971.

A. J. Callegari and J. B. Keller, "Contact of inflated membranes with rigid surfaces," J. Appl. Mech. 41, 189-191, 1974.

The existence and regularity theory for various types of free boundary problems is presented by:

A. Friedman, "Variational problems and free boundary problems," Wiley, New York, 1982.

The asymptotic solution presented in Section 6 was worked out for these notes.

# The Method of Partial Regularity as Applied to the Navier-Stokes Equations

By Robert V. Kohn
Courant Institute of Mathematical Sciences

The solutions of a system of partial differential equations are frequently studied in two steps: first one proves the existence of a weak solution in a suitable Sobolev space; then one proves the regularity of this weak solution. For nonlinear systems the second step may be too difficult--indeed, it may be false, the solutions may have singularities. In such cases one seeks a partial regularity theorem, restricting the size of the set of possible singularities, and one attempts the local description, to leading order, of the behavior near a singularity.

This approach has been used with great success in the theory of codimension one area-minimizing surfaces, developed by de Giorgi, Almgren, and others. There the weak solutions are described using integral currents or sets of finite perimeter; the singular set has codimension seven; and near any singular point the solution behaves locally like a minimal cone. The literature on this and other geometric applications is extensive; a rather accessible introduction is given in [3].

A similar approach has been used to study elliptic equations and systems. One recent success is the application by Schoen and Uhlenbeck to harmonic mappings [13]; other aspects of recent work are reviewed by Giaquinta and Giusti in [1].

The philosophy of partial regularity was extended beyond the realm of elliptic theory by V. Scheffer, in a highly original series of papers on the Navier-Stokes equations [7-12]. His analysis was recently extended and improved in [2], and that work will be the focus of this presentation. My real purpose is, however, not so much to explain these results as to describe the typical structure of a partial regularity theorem. The method is, in my opinion, a powerful one for which new applications will be found to nonlinear problems of all types--parabolic or hyperbolic, as well as elliptic or geometric. This article will be successful if it helps a few readers recognize new applications of partial regularity in problems of interest to them.

The main results on the Navier-Stokes equations which will be the medium for this discussion are these:

(0.1) The set of singular times has 1/2-dimensional measure zero in $\mathbb{R}_+$ [7].

(0.2) The set of singular points in space-time has one-dimensional measure zero in $\mathbb{R}^3 \times \mathbb{R}_+$ [2].

Each is proved by a combination of four ingredients:

A) an existence theory for weak solutions;

B) a scaling property of the underlying equations;

C) a local regularity result, which assumes that something is sufficiently small; and

D) a covering argument.

The existence theory serves, of course, to provide an object for study. Scaling allows one to "blow up" the solution near a possible singularity without changing the underlying equations. The local regularity result is the heart of the matter, and the most difficult of the four: it assures regularity if the "blown up" solution is not too large. Finally, the covering argument guarantees that (C) can be applied except on a rather small set.

## 1. Weak Solutions

For simplicity, I discuss only the homogeneous Navier-Stokes equations with unit viscosity on all of $\mathbb{R}^3$:

$$(1.1) \qquad \begin{aligned} u_t + u\cdot\nabla u - \Delta u + \nabla p &= 0 \\ \nabla\cdot u &= 0 \end{aligned}$$

Here $u:\mathbb{R}^3 \times \mathbb{R}_+ \to \mathbb{R}^3$ represents the fluid velocity, $p:\mathbb{R}^3 \times \mathbb{R}_+ \to \mathbb{R}$ is the pressure, and I am interested in the initial value problem

$$u\Big|_{t=0} = u_0(x).$$

Weak solutions were first studied by Leray in 1934. Following Scheffer and [2], I work here with the slightly stronger notion of a suitable weak solution: This means that

$$(1.2) \quad \sup_{0<t<\infty} \int_{\mathbb{R}^3 \times t} |u|^2 \, dx + \int_0^\infty \int_{\mathbb{R}^3} |\nabla u|^2 \, dxdt < \infty;$$

u and p satisfy (1.1) weakly in the usual sense; and for any smooth, nonnegative $\varphi$ compactly supported in space there is the "generalized energy inequality"

$$(1.3) \quad \int_{\mathbb{R}^3 \times t} |u|^2 \varphi dx + 2 \int_0^t \int_{\mathbb{R}^3} |\nabla u|^2 \varphi dxdt$$
$$\leqslant \int_{\mathbb{R}^3} |u_0|^2 \varphi dx$$
$$+ \int_0^t \int_{\mathbb{R}^3} [|u|^2(\varphi_t + \Delta\varphi) + (|u|^2 + 2p) \, u \cdot \nabla\varphi] \, dxdt$$

for each $t > 0$.

The generalized energy inequality follows formally (with =, not $\leqslant$ ) from (1.1) by multiplication with $u^i\varphi$, summation over i, and integration by parts. The bounds (1.2) yield $L^p$ estimates for both u and p: taking the divergence of (1.1) leads to

$$\Delta p = - \sum_{i,j=1}^{3} \frac{\partial}{\partial x_i \partial x_j} (u^i u^j)$$

at each time, so the pressure is given by a sum of singular integral operators acting on $u^i u^j$. This, (1.2), and standard interpolation inequalities assure that

(1.4) $$\int_0^T \int_{\mathbb{R}^3} (|u|^{10/3} + |p|^{5/3})\, dxdt < \infty \text{ for every } T > 0.$$

If $u_0$ is smooth, then so is $u(x,t)$ for a certain time interval; but (for large data) the only global existence theorems known are for weak solutions. It remains an open question whether a smooth solution can develop singularities in finite time. Other, presumably easier questions can be phrased as well: are weak solutions unique? And does the generalized energy inequality hold with equality? The partial regularity theory summarized here falls far short of answering these questions, but it does give some insight into the nature of any possible singularities.

## 2. Scaling

If u and p solve the Navier-Stokes equations,then so do

(2.1) $$u_\lambda(x,t) = \lambda u(\lambda x,\lambda^2 t) \quad \text{and} \quad p_\lambda(x,t) = \lambda^2 p(\lambda x,\lambda^2 t)$$

for each $\lambda>0$. If u is singular at (0,0) then each $u_\lambda$ is singular too, and studying $u_\lambda$ as $\lambda \to 0$ amounts to "blowing up" the singularity.

In view of this scaling, local statements about (1.1) are properly formulated on parabolic cylinders

(2.2) $$Q_r(x,t) = \{(y,\tau): \ |y-x| < r, \ |t-\tau| < r^2\}$$

rather than on Euclidean balls. An integral estimate on $Q_1(x,t)$ rescales to one on $Q_\lambda(x,t)$, for example

(2.3) $$\iint_{Q_\lambda(x,t)} (|u|^{10/3} + |p|^{5/3}) = \lambda^{5/3} \iint_{Q_1(x,t)} (|u_\lambda|^{10/3} + |p_\lambda|^{5/3})$$

$$\iint_{Q_\lambda(x,t)} |\nabla u|^2 = \lambda \iint_{Q_1(x,t)} |\nabla u_\lambda|^2.$$

It is convenient to assign a "scaling dimension" to each quantity:

$x_i$ has dimension 1 $\qquad$ $u^i$ has dimension -1

t has dimension 2 $\qquad$ p has dimension -2.

Then (2.3) can be summarized by saying that

$$r^{-5/3} \iint_{Q_r(x,t)} (|u|^{10/3} + |p|^{5/3}) \quad \text{and} \quad r^{-1} \iint_{Q_r(x,t)} |\nabla u|^2$$

have scaling dimension zero. Clearly, a scale-invariant local estimate must involve only zero-dimensional quantities.

## 3. The Set of Singular Times

The result (0.1) on the set of singular times is relatively easy, requiring little more than a restatement of Leray's early analysis. I call a time t regular if

(3.1) $$\sup_{t-\epsilon<\tau<t+\epsilon} \int_{\mathbb{R}^3} |\nabla u(x,\tau)|^2 \, dx < \infty$$

for some $\epsilon>0$; well-known higher regularity results assure that in that case u is $C^\infty$ on $\mathbb{R}^3$ x $(t-\epsilon,t+\epsilon)$. A time is singular if it is not regular, and the set of singular times will be denoted by $\Sigma$.

The statement that $\Sigma$ has "1/2-dimensional measure zero" means this: for every $\delta>0$ there is a cover of $\Sigma$ by intervals $\{(\sigma_i,\tau_i)\}$ such that

(3.2) $$\sum_{i=1}^{\infty} (\tau_i-\sigma_i)^{1/2} < \delta.$$

The heart of the proof is a local-in-time regularity result for "sufficiently small" solutions:

Proposition 1: There are absolute constants $\epsilon_1$, $C_1>0$ such that for any r, $0<r\leq t^{1/2}$,

$$(3.3)\quad r^{-1}\int_{t-r^2}^{t+r^2}\int_{\mathbb{R}^3} |\nabla u|^2\, dx d\tau < \epsilon_1 \Rightarrow$$

$$\sup_{t-\frac{r^2}{2}<\tau<t+\frac{r^2}{2}} r\int_{\mathbb{R}^3\times\tau} |\nabla u|^2\, dx < C_1.$$

I explain how this follows from Leray's estimates. By translation and scaling, it suffices to consider t=r=1, <u>i.e.</u> to show that

$$(3.3)'\quad \int_0^2\int_{\mathbb{R}^3} |\nabla u|^2 < \epsilon_1 \Rightarrow \sup_{\frac{1}{2}<\tau<\frac{3}{2}} \int_{\mathbb{R}^3\times\tau} |\nabla u|^2 < C_1.$$

If the hypothesis holds, then for some $t_0 \in (0,1/2)$

$$\int_{\mathbb{R}^3\times t_0} |\nabla u|^2\, dx < 2\epsilon_1.$$

Leray showed that the first singular time $t_1$ after $t_0$ satisfies

$$(t_1-t_0) > C\left[\int_{\mathbb{R}^3\times t_0} |\nabla u|^2\, dx\right]^{-2};$$

if $\epsilon_1$ is small enough then $t_1>3/2$, and the conclusion of (3.3)' follows.

In the form (3.3)', Proposition 1 says that a sufficiently small weak solution on a unit-sized time interval (0,2) is regular on the subinterval (1/2,3/2). The scaled version (3.3) looks a bit different, however: it applies whenever the average of $\int |\nabla u|^2\, dx$ over the interval $(t-r^2,t+r^2)$ is less than $\epsilon_1 r^{-1}$. Since $\epsilon_1 r^{-1} \to \infty$ as $r \to 0$, (3.3) prescribes a <u>minimum</u> <u>rate</u> at which singularities can develop.

Estimating the size of $\Sigma$ involves a covering argument based on this simple observation: if t is a singular time then the hypothesis of (3.3) must fail for every r, so that

$$\int_{t-r^2}^{t+r^2} \int_{\mathbb{R}^3} |\nabla u|^2 \, dx d\tau \geq \epsilon_1 r \text{ for } 0<r\leq t^{1/2}.$$

By the Vitali covering lemma, for any neighborhood N of $\Sigma$ there is a disjoint family of intervals $\{(t_i-\rho_i,\ t_i+\rho_i)\}$ contained in N such that

$$(3.4) \quad \int_{t_i-\rho_i}^{t_i+\rho_i} \int_{\mathbb{R}^3} |\nabla u|^2 \, dx d\tau \geq \epsilon_1 \rho_i^{1/2} \quad \text{for each } i$$

and

$$(3.5) \quad \Sigma \subset \bigcup_{i=1}^{\infty} (t_i-5\,\rho_i,\ t_i+5\,\rho_i).$$

Addition of (3.4) gives

$$(3.6) \quad \epsilon_1 \sum_{i=1}^{\infty} \rho_i^{1/2} \leq \int_{N \times \mathbb{R}^3} |\nabla u|^2 \, dxdt < \infty,$$

making use of (1.2). This shows that $\Sigma$ has Lebesque measure zero, and it follows that the right side of (3.6) can be made as small as desired by an appropriate choice of N. Thus, $\Sigma$ has been covered by intervals $\{(t_i-5\,\rho_i,\ t_i+5\,\rho_i)\}$ with

$$(3.7) \quad \sum_{i=1}^{\infty} (5\,\rho_i)^{1/2} \leq \frac{\sqrt{5}}{\epsilon_1} \iint_{N \times \mathbb{R}^3} |\nabla u|^2 \, dxdt \to 0,$$

verifying (3.2).

## 4. The Singular Set in Space-Time

The size of the singular set will again be estimated by combining a dimensionless, local regularity result with an easy covering argument. This time a point (x,t) will be called "singular" if u is unbounded there--more precisely, the singular set is

$$S = \{(x,t):\ u \text{ is not } L^{\infty}_{loc} \text{ in any neighborhood of } (x,t)\}.$$

Serrin's higher regularity theory shows that u is $C^{\infty}$ in x (though not necessarily in t) off S.

In saying that S has "one-dimensional measure zero," I mean that for every $\delta>0$ there is a cover of S by parabolic cylinders $\{Q_{r_i}(x_i,t_i)\}$ such that

$$(4.1) \qquad \sum_{i=1}^{\infty} r_i < \delta.$$

This is the analogue in the parabolic metric of having one-dimensional Hausdorff measure zero. For any such S, its projection onto the t-axis has 1/2-dimensional measure zero; therefore, (0.2) actually includes (0.1) as a corollary (with a little fussing over the different notions of "regularity").

As before, the heart of the matter is a local regularity result:

<u>Proposition 2</u>: <u>There is an absolute constant $\epsilon_2>0$ such that</u>

$$(4.2) \qquad \limsup_{r\to 0} r^{-1} \iint_{Q_r(x,t)} |\nabla u|^2 < \epsilon_2 \Rightarrow (x,t) \notin S.$$

The hypothesis of (4.2) is again a statement about the rate at which $|\nabla u|^2$ blows up: roughly speaking, Proposition 2 says that near any singularity (x,t)

$$|\nabla u|^2(y,\tau) \geq \epsilon_2 \rho^{-4}, \qquad \text{where } \rho(y,\tau) = |y-x| + |\tau-t|^{1/2}.$$

The proof is rather long, and it will not be discussed here. An expository treatment is given in [5], and the details are in [2].

The covering argument leading from Proposition 2 to (0.2) is almost the same as in Section 3. If $(x,t) \in S$ then by (4.2), there exist arbitrarily small $r>0$ with

$$\iint_{Q_r(x,t)} |\nabla u|^2 \geqslant \epsilon_2 r.$$

The analogue of Vitali's covering Lemma is valid for parabolic cylinders; it yields a disjoint family $\{Q_{r_i}(x_i,t_i)\}$ contained in a neighborhood N of S such that

(4.3) $$\iint_{Q_{r_i}(x_i,t_i)} |\nabla u|^2 \geqslant \epsilon_2 r_i \qquad \text{for each } i$$

and

(4.4) $$S \subset \bigcup_{i=1}^{\infty} Q_{5r_i}(x_i,t_i).$$

Addition of (4.3) gives

(4.5) $$\epsilon_2 \sum_{i=1}^{\infty} r_i \leqslant \iint_N |\nabla u|^2 < \infty.$$

It follows as before that S has Lebesque measure zero, so that when N shrinks to S

(4.6) $$\sum_{i=1}^{\infty} r_i \leqslant \epsilon_2^{-1} \iint_N |\nabla u|^2 \to 0,$$

establishing (4.1).

A review of the preceding argument reveals that <u>the scaling dimension of the local regularity result determines the estimated dimension of the singular set</u>. To clarify this point, consider what happens if Proposition 2 is replaced by

<u>Proposition 3</u>: <u>There are universal constants</u> $\epsilon_3$, $C_3$ <u>such that for any</u> $r>0$

(4.7) $$r^{-5/3} \iint_{Q_r(x,t)} (|u|^{10/3} + |p|^{5/3}) < \epsilon_3 \Rightarrow |u| \leqslant C_3 r^{-1} \text{ a.e. on } Q_{r/2}(x,t).$$

This assertion, an easy corollary of results in [2], is a local version of Scheffer's estimate [9]. If used instead of Proposition 2 it yields

$$(4.3)' \quad \iint_{Q_{r_i}(x_i,t_i)} (|u|^{10/3} + |p|^{5/3}) \geq \epsilon_3 \, r_i^{5/3},$$

in place of (4.3) and

$$(4.6)' \quad \sum_{i=1}^{\infty} r_i^{5/3} \leq \epsilon_3^{-1} \iint_N (|u|^{10/3} + |p|^{5/3}) \to 0$$

in place of (4.6), using (1.4) rather than (1.2) in the last step. Thus, Proposition 3 leads to the weaker conclusion that S has 5/3-dimensional measure zero.

The key difference between (4.2) and (4.7) is the choice of integrand in the hypothesis: the former gives a one-dimensional result because $\iint |\nabla u|^2$ has scaling dimension one. The method clearly requires that the integrand--$|\nabla u|^2$ in (4.2), and $(|u|^{10/3} + |p|^{5/3})$ in (4.7)--be globally integrable. Since the basic estimates (1.2) for weak solutions involve one-dimensional quantities, it seems difficult to improve (0.2) by the present method without a new global estimate having scaling dimension less than one.

## 5. Conclusion

I have sketched the role of the four basic ingredients--global estimates, scaling, local regularity, and covering--in the particular context of the Navier-Stokes equations. It should, however, be evident that this is a rather flexible approach, potentially of use in many different problems. The hardest step is always the local regularity: its form and proof are highly problem-dependent.

The story of the Navier-Stokes equations remains far from complete: the big question is whether or not singularities exist at all. My personal opinion is that they do, and that Leray's self-similar ansatz proposed in [6] correctly describes their local behavior. Though this remains at present a distant goal, results of a similar character have recently been obtained with Y. Giga for the far simpler model problem $u_t - \Delta u - u^p = 0$ [4].

References


[1] Ball, J. M., ed., Systems of Nonlinear Partial Differential Equations, D. Reidel Pub. Co., Dordrecht, 1983.

[2] Caffarelli, L., Kohn, R., and Nirenberg, L., Partial Regularity of Suitable Weak Solutions of the Navier-Stokes Equations, C.P.A.M. 35, 771-831 (1982).

[3] Giusti, E., Minimal Surfaces and Functions of Bounded Variation, Australia Nat'l. Univ. Notes in Pure Math 10 Canberra, 1977.

[4] Giga, Y., and Kohn, R., Asymptotically Self-Similar Blow-up of Semilinear Heat Equations, to appear.

[5] Kohn, R., Partial Regularity and the Navier-Stokes Equations, in Nonlinear Partial Differential Equations in Applied Science, H. Fujita, P. D. Lax, and G. Strang, eds., North Holland Math. Studies No. 81, 1983, 101-118.

[6] Leray, J., Sur le mouvement d'un liquide visquex emplissant l'espace, Acta Math. 63, 193-248 (1934).

[7] Scheffer, V., Turbulence and Hausdorff Dimension, in Turbulence and the Navier-Stokes Equations, Lecture Notes in Math 565, Springer-Verlag, 1976, 94-112.

[8] ------------------, Partial Regularity of Solutions to the Navier-Stokes Equations, Pacific J. Math 66, 535-552 (1976).

[9] --------------------, Hausdorff Measure and the Navier-Stokes Equations, Comm. Math. Phys. 55, 97-112 (1977).

[10] ----------------, The Navier-Stokes Equations in Space Dimension Four, Comm. Math. Phys. 61, 41-68 (1978).

[11] ------------------, The Navier-Stokes Equations on a Bounded Domain, Comm. Math. Phys. 73, 1-42 (1980).

[12] -----------------, Boundary Regularity for the Navier-Stokes Equations in a Half-Space, Comm. Math. Phys. 85, 275-299 (1982).

[13] Schoen, R., and Uhlenbeck, K., A Regularity Theory for Harmonic Maps, J. Diff. Geometry 17 (1982), 307-335 and ibid. 18 (1983) 329.

# Shock Waves, Increase of Entropy and Loss of Information

By Peter D. Lax[1]

## 1. Introduction

We present an informal review of the topics in the title as they pertain to solutions of hyperbolic systems of conservation laws. These are systems of the form

$$(1.1)\qquad u^i_t + f^i_x = 0,\ i = 1, \ldots, n;$$

the subscripts t and x denote partial derivatives. Each $u^i$ is a density, $f^i$ the corresponding flux. Each $f^i$ is a function of all $u^j$'s, so $f^i_x$ can be expressed as a linear combination of $u^j_x$. In matrix notation (1.1) can be written as

$$(1.2)\qquad u_t + Au_x = 0$$

where u denotes the column vector with components $u^j$, and A the matrix whose $i^{th}$ row is the u-gradient of $f^i$:

$$(1.3)\qquad a_{ij} = \frac{\partial f^i}{\partial u^j}$$

The matrix A is a function of u, unless $f^i$ are linear functions of u; in this talk we deal with systems that are genuinely nonlinear, in a sense to be made precise.

---

[1]Research supported under contract DE-AC02-76ER03077 of the Applied Mathematical Sciences Research Program of the Office of Basic Energy Sciences, DOE.

The system (1.2) is called <u>strictly hyperbolic</u> if the matrix A has real and distinct eigenvalues $\alpha_1, \alpha_2, \ldots, \alpha_n$ for all values of u.

We are interested in solving the initial value problem:

(1.4) $u(x,0) = u_0(x)$,

in particular, we want to study the nature of the dependence of solutions on their initial data. We denote by S(t) the operator relating solutions at time t to their initial values:

$$S(t):u(x,0) \rightarrow u(x,t).$$

The main facts of life :

(i) The initial value problem has no proper, i.e., differentiable solution for all time t, no matter how smooth the initial data are.

(ii) The initial value problem can be solved for all time if we admit solutions of (1.1) in the integral sense, i.e., in the sense of distributions.

Solutions in the distribution sense that are piecewise continuous satisfy the Rankine-Hugoniot relation

(1.5) $s[u^i] = [f^i],\ i = 1, \ldots, n,$

where [ ] denotes the difference between the value on the left side and the right side of the discontinuity; s is the velocity with which the discontinuity propagates.

(iii) Solutions in the distributions sense are not determined uniquely by their initial data.

In these notes we shall describe various criteria that are used to accept or reject distribution solutions. The criteria are suggested by physical facts, and are analyzed mathematically. The analysis shows, or at least leads one to expect,

(iv) The various criteria all pick out the same class of distribution solutions; we shall call these relevant solutions.

(v) Each member of the class of relevant solutions is uniquely determined by its initial data. The initial data can be prescribed arbitrarily within the class of $L^{\infty}$ functions.

We denote by S(t) the operator linking relevant solutions to their initial data.

The main theme of this talk is the following remarkable property of relevant solutions:

(vi) The set of relevant solution is compact; in particular, the operators S(t), $t>0$, are compact in appropriate pairs of topologies.

Note that (vi) is a nonlinear property; for linear hyperbolic equations S(t) is invertible, in most cases unitary. In the rest of these notes we shall explore how nonlinearity brings about compactness.

Very little is known about the compactness of S(t) in more than one space dimension.

The organization of these notes is as follows:

In Section 2 we discuss, for the simplified model of a single conservation law, the concepts of genuine nonlinearity, breakdown of classical solutions, solutions in the distribution sense and their nonuniqueness, the viscosity method, finite difference methods, and the shock condition.

In Section 3 we discuss, for the scalar model, the compactness of solutions constructed by the viscosity and difference methods, and derive the entropy inequality for such solutions.

In Section 4 we derive Glimm's estimate for the total variation of solutions of scalar equations that satisfy the shock condition, we show that a discontinuous solution that satisfies the shock condition also satisfies the entropy condition.

In Section 5 we indicate how to extend the notions developed in Sections 2, 3, and 4 for systems.

Section 6 contains scattered remarks about the equations of compressible flow: the increase of entropy, some consequences of Carnot's theorem, and the equipartition of energy in the wake of strong shocks.

Compactness of the family of all solutions places a limit on the amount of information contained in that family. It is natural to ask what this implies about the amount of computing needed to solve an initial value problem to meet a specified degree of resolution; see [14] for some crude notions.

[13] contains a bird's eye view of conservation laws; [16] is a thorough treatise.

## 2. Single Conservation Laws

We study (1.1) for $n = 1$:

$$u_t + f_x = 0, \tag{2.1}$$

$f$ some function of $u$; we denote

$$\frac{df}{du} = a(u). \tag{2.2}$$

As in (1.2), Equation (2.1) can be rewritten as

$$u_t + a\, u_x = 0; \tag{2.3}$$

this equation asserts that $u$ is constant along trajectories $x = x(t)$ satisfying

$$\frac{dx}{dt} = a. \tag{2.4}$$

In view of this interpretation, $a$ is called the signal velocity; the trajectories defined by (2.4) are called characteristics. Since according to (2.3), $u$ is constant along characteristics, and since $a$ is a function of $u$, it follows from (2.4) that the characteristics are straight lines.

When the initial value $u_0$ of u is specified, we can construct through each point (y,0) on the initial line a characteristic line:

(2.5) $x = y + a(u_0(y))t.$

Suppose $y_1$ and $y_2$ are two points, $y_1<y_2$, and the inequality

(2.6) $a(u_1)>a(u_2)$,

holds, where

(2.6)' $u_1 = u_0(y_1)$, $u_2 = u_0(y_2)$.

(2.5) shows that the two characteristics issuing from $(y_1,0)$, respectively, intersect at the time

$$T = \frac{y_2 - y_1}{a(u_1) - a(u_2)}$$

As we saw before, u has the value $u_1$ along the whole characteristic issuing from $(y_1,0)$, and the value $u_2$ along the characteristics issuing from $(y_2,0)$. So at the point of intersection u has to be equal to both $u_1$ and $u_2$; since (2.6) shows that $u_1 \neq u_2$, this is impossible and shows that no differentiable solution u(x,t) exists beyond the time T.

Note that the crucial inequality (2.6) can hold only if a(u) is a genuine function of u, which makes f(u) a genuinely nonlinear function of u. It is convenient to assume that $\frac{d}{du} a \neq 0$ for all u; in view of (2.2) this implies that

(2.7) $$\frac{d^2 f}{du^2} \neq 0,$$

i.e., that f is strictly convex or concave.

As noted in the introduction, solutions that cannot be continued beyond some critical time T can, nevertheless, be contrived as solutions of (2.1) in the sense of distributions:

(2.8) $$\iint_{t\geq 0}(u\, w_t + f\, w_x)\, dxdt + \int u_0\, w_0 dx = 0$$

for every test function w in $C_0^\infty$. This is consistent with the point of view of physics that conservation laws are integral relations:

(2.8)' $$\frac{d}{dt}\int_G u dx + \int_{\partial G} f \cdot n dS = 0$$

for every domain G. Relations (2.8) can be deduced from relations (2.8) by a simple process of approximation.

As remarked in Section 1, a piecewise continuous solution is a solution in the distribution sense if the Rankine-Hugoniot relation (1.5) is satisfied across the discontinuity;

(2.9) $$s[u] = [f],$$

where [ ] denotes the difference across the discontinuity, and s is the velocity with which the discontinuity propagates.

A simple example of such a discontinuous solution of Equation (2.1), with

(2.10) $$f(u) = u^2/2$$

is

(2.11) $$u_1(x,t) = \begin{cases} 1 & \text{for } x < t/2 \\ 0 & \text{for } t/2 < x. \end{cases}$$

The discontinuity is across the line

(2.12) $$x = t/2$$

which propagates with velocity $s = 1/2$; across the discontinuity $[u] = 1$ $[f] = 1/2$, so relation (2.9) is satisfied.

The function

$$(2.13) \quad u_2(x,t) = \begin{cases} 0 & \text{for } x < t/2 \\ 1 & \text{for } t/2 < x \end{cases}$$

also is discontinuous across the line (2.12), and $[u] = -1$, $[f] = -1/2$, so that relation (2.9) is satisfied. Note, however, that the function

$$(2.14) \quad u_3(x,t) = \begin{cases} 0 & \text{for } x \leqslant 0 \\ x/t & \text{for } 0 \leqslant x \leqslant t \\ 1 & \text{for } t \leqslant x \end{cases}$$

satisfies Equation (2.1) with f given by (2.10), so that

$$(2.15) \quad u_t + uu_x = 0,$$

in each of the three regions, and is <u>continuous</u> across the boundaries $x = 0$ and $x = t$ separating the regions. Since $u_2(x,0) \equiv u_3(x,0)$, solutions in the distribution sense are <u>not uniquely determined by their initial data</u>. This shows that an additional criterion is needed, based on physical principles and buttressed by mathematical ones, which rejects certain distribution solutions. The remaining acceptable ones must have the property that every initial value problem has exactly one acceptable solution.

There are several ways of formulating such a criterion of rejection or acceptance; happily, they turn out to be equivalent. We list the most important ones:

(a) The acceptable solutions u of (2.1) are the limits of solutions $u^{(\epsilon)}$ of a family of equations obtained by augmenting the flux f by a small viscous term, and letting the viscosity tend to 0:

$$(2.16) \quad u = \lim_{\epsilon \to 0} u^{(\epsilon)}.$$

The viscous term in the flux is $-\epsilon u_x$, $\epsilon>0$, so that $u^{(\epsilon)}$ satisfies the equation

$$(2.17) \quad u_t^{(\epsilon)} + f_x^{(\epsilon)} = \epsilon u_{xx}^{(\epsilon)}, \qquad f^{(\epsilon)} = f(u^{(\epsilon)}),$$

and has the same initial value as u:

$$(2.18) \quad u^{(\epsilon)}(x,0) = u_0(x).$$

(b) The acceptable solutions are limits of solutions v of a difference approximation as $\Delta t$, $\Delta x$ tend to zero. Denoting the value of v at $x = k\Delta x$, $t = n\Delta t$ by $v_k^n$, the difference equation is of the form

$$(2.19) \quad v_k^{n+1} = \frac{1}{2}\left[u_{k+1}^n + u_{k-1}^n\right] - \left[f_{k+1}^n - f_{k-1}^n\right]\frac{\Delta t}{2\Delta x};$$

here we use the abbreviation

$$f_j^n = f\left[u_j^n\right].$$

The initial values are

$$v_k^0 = u_0(k\Delta x).$$

The rationale for using limits of (2.19) is the close relation between (2.19) and (2.17); this can be seen by using the Taylor approximations

$$v_k^{n+1} = u + \Delta t\, u_t + \frac{1}{2}(\Delta t)^2 u_{tt} + 0(\Delta^3)$$

$$\frac{1}{2}\left[u_{k+1}^n + u_{k-1}^n\right] = u + \frac{1}{2}(\Delta x)^2 u_{xx} + 0(\Delta^3)$$

$$\frac{1}{2}\left[f_{k+1}^n - f_{k-1}^n\right] = \Delta x\, f_x + 0(\Delta^3).$$

Setting these into (2.19) and using (2.1) to calculate the higher derivatives of u we get that v approximates solutions of (2.17), with

$$(2.20) \quad \epsilon = \frac{\Delta t}{2}\left[\left(\frac{\Delta x}{\Delta t}\right)^2 - a^2\right]$$

Since $\epsilon$ has to be positive, we must have

$$(2.21) \quad \frac{\Delta x}{\Delta t} \geq |a|;$$

this is the celebrated Courant-Friedricks-Lewy convergence criterion.

We remark that (2.19) is merely one of a variety of difference approximations we may use. Another class of approximations, combining viscosity in space and discretization in time, is Avron Douglis' layering method; it is more flexible than either method (a) or (b), see [3].

An entirely different criterion for accepting or rejecting distribution solutions can be based on the analysis of the mechanism that causes the breakdown of smooth solutions: the intersection of characteristics. Thus, the characteristics issuing from the initial line for the initial data of $u_1$, given by (2.11), cross in a wedge-shaped region

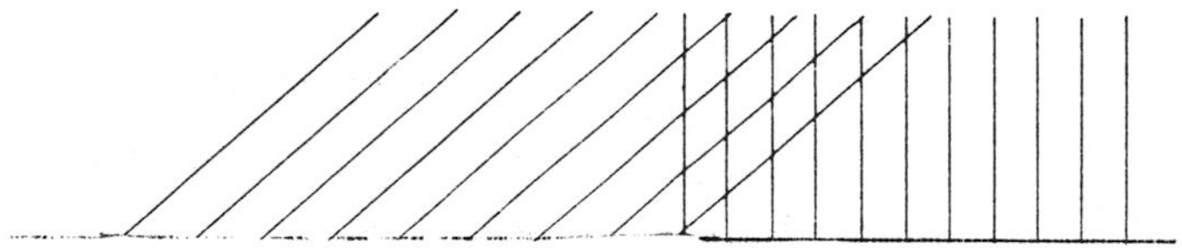

Fig. 1

The role of the discontinuity in (2.11) along $x = t/2$ is to keep the characteristics from crossing. This is in contrast to the behavior of the characteristics issuing from the initial line for the data of $u_2$, given by (2.13); these diverge and don't cross at all:

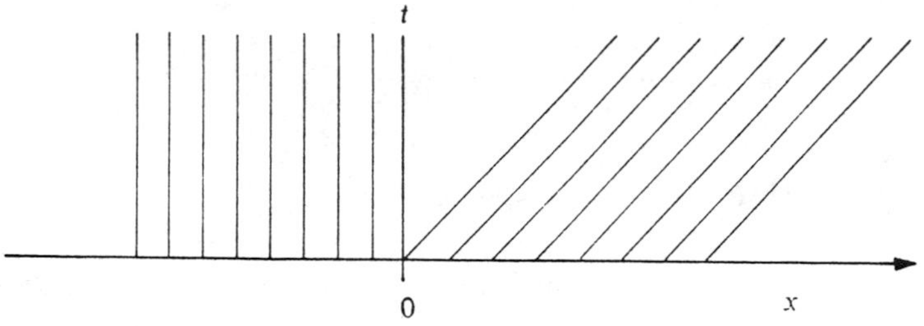


Fig. 2

This leads to the concept of a shock:

A discontinuity of a piecewise continuous distribution solution is called a shock if the characteristics on either side impinge in the forward t direction on the discontinuity. Denoting by $u_L$ and $u_R$ the value of u on the left and right sides of the discontinuity, and by $a_L$ and $a_R$ the corresponding signal velocities:

(2.22) $a_L = a(u_L),\ a_R = a(u_R),$

we can express this condition by the inequality

(2.23) $a_L > s > a_R,$

where s is the velocity of propagation of the discontinuity. We recall from (2.9) that

(2.24) $s = \dfrac{f_R - f_L}{u_R - u_L},$

where

(2.25) $f_L = f(u_L)$, $f_R = f(u_R)$.

(c) A distribution solution of (2.1) is acceptable if all its discontinuities are shocks, i.e., satisfy condition (2.23).

## 3. Viscosity Methods

As a start, let's assume that for arbitrary bounded measurable initial data $u_0$ of compact support, the solutions $u^{(\epsilon)}(x,t)$ of (2.17), (2.18) exist and converge in the $L^1(dxdt)$ norm over $\mathbb{R} \times (0,T)$, T any value > 0. What can we deduce about the limit u?

Multiplying equation (2.17) by any $C_0^2$ test function w(x,t) and integrating by parts over all x and $t \geqslant 0$; we get

$$(3.1)\quad -\iint \left[u^{(\epsilon)}w_t + f(u^{\epsilon})w_x\right] dxdt - \int u_0 w_0 dx = \epsilon \iint u\ w_{xx}\ dxdt.$$

As $\epsilon$ tends to zero, $u^{(\epsilon)}$ tends to u and $f(u^{(\epsilon)})$ tends to f(u) in the $L^1$ norm. Therefore, (3.1) tends to relation (2.8), characterizing distribution solutions; this shows that the $L^1$ limit of solutions $u^{(\epsilon)}$ of (2.17) is a solution of (2.1) in the distribution sense.

Let $u^{(\epsilon)}$ be a solution of (2.17); since the initial data $u_0(x)$ has compact support, $u^{(\epsilon)}(x,t)$ tends to 0 rapidly as $|x| \to \infty$. So the function

$$(3.2)\quad U^{(\epsilon)}(x,t) = \int_{-\infty}^{x} u^{(\epsilon)}(y,t)dy$$

is well defined and bounded. Clearly,

$$(3.2)'\quad U_x^{(\epsilon)} = u^{(\epsilon)}.$$

Integrating (2.17) with respect to x gives

$$(3.3)\quad U^{(\epsilon)} + f\left[U_x^{(\epsilon)}\right] = \epsilon U_{xx}^{(\epsilon)},$$

provided that the flux is normalized so that

(3.4) $f(0) = 0.$

Let $v^{(\epsilon)}$ be another solution of (2.17), $V^{(\epsilon)}$ its x-integral, satisfying an analogue of (3.3). Subtracting the two equations from each other, and applying the mean value theorem, we deduce that the difference

(3.5) $D = U^{(\epsilon)} - V^{(\epsilon)}$

satisfies

(3.6) $D_t + \bar{a}D_x = \epsilon D_{xx}$,

where

$$\bar{a} = \frac{f(u^{\epsilon}) - f(v^{\epsilon})}{u^{\epsilon} - v^{\epsilon}}$$

We apply now the <u>maximum principle</u> to the parabolic equation (3.6) and deduce that

(3.7) $|D|_{max}(t) = \operatorname{Max}_x |D(x,t)|$

is a <u>decreasing</u> function of t.

Clearly,

$$|U^{(\epsilon)}(t)|_{max} \leq \int_{\mathbb{R}} |u^{(\epsilon)}(x,t)|\, dt,$$

so

(3.8) $\int_0^T |U^{(\epsilon)}|_{max}(t)\, dt \leq \int_0^T\int_{\mathbb{R}} |u^{(\epsilon)}(x,t)|\, dxdt$

By assumption, $u^{(\epsilon)}$ converges to u in the $L^1$ norm; it follows, therefore, from (3.8) that $U^{(\epsilon)}$ converges to U in the $L^1$(Max) norm on the left of (3.8). Similarly, $V^{(\epsilon)}$ converges to V; by the triangle inequality, for each t

$$\left| |U^{(\epsilon)}-V^{(\epsilon)}|_{max} - |U-V|_{max} \right| \leq |U^{(\epsilon)}-U|_{max} + |V^{(\epsilon)}-V|_{max}$$

It follows from this that $|U^{(\epsilon)} - V^{\epsilon}|_{max}(t)$ tends to $|U - V|_{max}(t)$ in the $L^1(dt)$ norm. Since by (3.5) and (3.7) the former is a decreasing function of t, and since the $L^1$ limits of decreasing functions are decreasing, it follows that

$$(3.9) \qquad |U - V|_{max}(t)$$

is a decreasing function of t.

The quantity $|U|_{max}$ is called the $W^{-1,\infty}$ norm u; so property (3.9) can be expressed in terms of the solution operator S(t) relating initial data to data at time t:

<u>Theorem 3.1</u>: <u>The operators</u> S(t) <u>are contractions in the</u> $W^{-1,\infty}$ <u>norm</u>.

Since equation (2.17) is parabolic, another application of the maximum principle shows that

$$\operatorname*{Max}_x |u^{(\epsilon)}(x,t)| = |u^{(\epsilon)}|_{max}(t)$$

is a decreasing function of t. We conclude, as before, that $|u^{\epsilon}|_{max}(t)$ tends to $|u|_{max}(t)$ in the $L^1$ norm, and, therefore, that the latter also is a decreasing function of time. This property can be expressed so:

Theorem 3.2: The operators S(t) map into itself any ball in $L^\infty$ centered at 0.

Theorems 3.1 and 3.2 derived in [8]; it was observed there that they have the following surprising consequence:

Theorem 3.3: Suppose equation (2.1) is genuinely nonlinear, i.e., the function f is strictly convex or concave. Then the operators S(t), t>0, map any bounded subset of $L^\infty$ supported on a given interval into a compact subset of $L^1$.

Proof: Let $u_0^{(n)}$ be a uniformly bounded sequence of functions supported on a common interval of the x-axis. Such a set belongs to a compact subset of $W^{-1,\infty}$, i.e., a subsequence of $u_0^{(n)}$ converges to a limit $u_0$ in the $W^{-1,\infty}$ norm. This limit is also bounded and has compact support. According to Theorem 3.1, for each t,$u^{(n)}(x,t)$ converges in the $W^{-1,\infty}$ norm to u(x,t), where $u^{(n)}$ and u are the solutions of (2.1) constructed by the viscosity method with initial values $u_0^{(n)}$ and $u_0$, respectively. Both $u^{(n)}$ and u satisfy Equation (2.1) in the distribution sense (2.8): for any $C_0^\infty$ test function w,

$$(3.10)\quad \iint \left[u^{(n)}w_t + f(u^{(n)})w_x\right] dxdt + \int u_0^{(n)} w_0 dx = 0,$$

and

$$(3.10)'\quad \iint (u\, w_t + f(u)w_x)\, dxdt + \int u_0 w_0 dx = 0$$

Since $u^{(n)}$ tends to u in the $W^{-1,\infty}$ norm, it follows that the first and the third term in (3.10) tends to the first and third term in (3.10)'. Therefore, it follows that so does the second term:

$$(3.11)\quad \iint \left[f(u^{(n)}) - f(u)\right] w_x\, dxdt \to 0$$

Next we make use of the fact that not only for genuine solutions but also for distribution solutions, signals propagate with finite speed.

Since $u_0^{(n)} = u_0 = 0$ outside some finite x-interval, it follows that $u^{(n)}$ and u are zero outside a bounded act in x,t space for $0 \leq t \leq T$. We choose w so that $w_x \equiv 1$ on this bounded set; since $f(u^n) = f(u)$ outside this set, we can rewrite (3.11) as

$$(3.11)' \quad \iint \left[f(u^{(n)}) - f(u)\right] dxdt \to 0.$$

We have assumed that f is strictly, convex, say $f''(u) \geq R > 0$. Then

$$(3.12) \quad f(u^{(n)}) - f(u) \geq f'(u)(u^{(n)} - u) + \frac{R}{2}(u^{(n)} - u)^2$$

Integrating this we get

$$(3.12)' \quad \iint \left[f(u^{(n)}) - f(u)\right] dxdt \geq \iint \left[f'(u)\,(u^{(n)} - u)\right] dxdt$$
$$+ \frac{R}{2} \iint \left[u^{(n)} - u\right]^2 dxdt$$

Since $u^{(n)} - u$ tends to zero in the $W^{-1,\infty}$ norm, and since $u^{(n)} - u$ is uniformly bounded, it follows readily, by approximating $f'(u)$ in the $L^1$ sense by smooth functions, that the first term on the right in (3.12)' tends to zero. Since by (3.11)' the left side tends to zero, we deduce from (3.12)' that

$$(3.13) \quad \int_0^T \int (u^{(n)} - u)^2 \, dxdt \to 0.$$

Since $u^{(n)} - u$ is supported on a bounded set, it follows that

$$(3.13)' \quad \int^T \int |u^{(n)} - u| \, dxdt \to 0.$$

This proves $L^1(dxdt)$ convergence of $u^{(n)}$.

To prove $L^1(dx)$ convergence, we appeal to the following theorem of Barbara Keyfitz [7]:

If u and v are two distribution solutions of (2.1) that are limits of solutions of (2.17), then

$$\int |v(x,t) - u(x,t)| \, dx$$

is a decreasing function of time. This result applies in particular to $v = u^{(n)}$; from this and (3.13)' we conclude that

$$\int |u^{(n)}(x,t) - u(x,t)| \, dx \to 0$$

for each $t>0$. This completes the proof of Theorem 3.3.

It was already observed in [8] that Theorems 3.1, 3.2 and 3.3 hold also for solutions constructed by the difference scheme (2.19); the same is true when solutions are constructed by more general difference schemes, as long as they are of monotone type.

A basic hypothesis of Theorems 3.1–3.3 is that the solutions $u^{(\epsilon)}$ of the parabolic equation (2.17) converge to a limit u in the $L^1$ norm for all bounded initial data $u_0$. How does one prove such a result? In the case (2.10) of a quadratically nonlinear f, E. Hopf, used the fact that in this case Equation (3.3) is changed by the transformation

$$U^{(\epsilon)} = -2\epsilon \log V^{(\epsilon)}$$

into

$$V_t^{(\epsilon)} = \epsilon V_{xx}^{(\epsilon)}, \quad V(x,0) = \exp\{U_0^{(\epsilon)}/2\epsilon\}$$

A solution of the heat equation can be expressed as an integral of its initial values. Using this formula, Hopf was able to show that as $\epsilon \to 0$, $u^{(\epsilon)}$ tends to a limit u; he even obtained a fairly explicit formula for this limit. It was remarked in [9] that a version of Theorem 3.3 can be derived from this formula.

The convergence of solutions of the difference scheme (2.19) can be proved in a similar fashion for the special choice $f(u) = \log(a + be^{-u})$, and of other monotonic schemes for other special choices.

Needless to say, these methods are very special. Recently, see [15], [17] a new method, capable of dealing with more general cases, has been introduced by L. Tartar and F. Murat; we give a brief description of their ideas.

The first idea, going back to L. C. Young, is a precise description of weak convergence. Let $u^{(\epsilon)}(y)$ be a uniformly bounded sequence of functions; then it has a subsequence with the property that for every continuous function g the weak$^*$ limits in the sense of $L^\infty$ exist:

$$(3.14) \quad g(u^{(\epsilon)}) \rightharpoonup u_g.$$

Clearly, the weak$^*$ limits $u_g$ depend linearly and positively on g. It is not hard to show that $u_g$ can be represented as an integral of g with respect to a family of probability measures $\nu(y)$:

$$(3.15) \quad u_g(y) = \int g(v)\, d\nu(v,y) \equiv \langle g, \nu(y) \rangle.$$

Secondly, it is not hard to show, see (3.12)', that (3.14) is convergence in the $L^1$ sense if the measure $\nu(y)$ is concentrated at a single point for almost all values of y. In this case

$$(3.16) \quad u_g = g(u)$$

for every g, u being the limit of $u^{(\epsilon)}$.

Tartar takes for $u^{(\epsilon)}$ the solutions of (2.17) with prescribed initial values. By Theorem 3.2 this sequence is uniformly bounded; therefore, it has a subsequence for which (3.14), (3.15) holds. Tartar shows that in this case $\nu$ is concentrated at a single point by using the notion of

Compensated Compactness (Tartar):

Let $h^{(\epsilon)}$ and $k^{(\epsilon)}$ be two sequences of vector functions defined in some domain of y-space, satisfying the following conditions:

(i) $h^{(\epsilon)}$ and $k^{(\epsilon)}$ are uniformly bounded in $L^2$, and converge weakly in $L^2$:

(3.17) $h^{(\epsilon)} \rightharpoonup h,\ k^{(\epsilon)} \rightharpoonup k.$

(ii) $-\mathrm{div}\ h^{(\epsilon)}$ and curl $k^{(\epsilon)}$ belong to compact subsets of $H^{-1}_{loc}$.

<u>Conclusion</u>: The scalar product of $h^{(\epsilon)}$ and $k^{(\epsilon)}$ tend in the distribution sense to

(3.18) $h^{(\epsilon)} \cdot k^{(\epsilon)} \rightharpoonup h \cdot k.$

Take y to be t, x, the domain to be $(0,T) \times [x_1,x_2]$ and the vector function $h^{(\epsilon)}$ to be

(3.19) $h^{(\epsilon)} = (u^{(\epsilon)}, f^{\epsilon}),$

where $f^{(\epsilon)} = f(u^{\epsilon})$. Using Equation (2.17) we see that

(3.20) $\mathrm{div}\ h^{(\epsilon)} = u^{(\epsilon)}_t + f^{(\epsilon)}_x = \epsilon\, u^{\epsilon}_{xx}$

Clearly, the $H^{-1}_{loc}$ norm of div $h^{(\epsilon)}$ is bounded by $\epsilon \|u^{\epsilon}_x\|$, where $\|\ \|$ denotes $L^2$ norm. This quantity can be estimated by multiplying (2.17) by $u^{(\epsilon)}$ and integrating over $\mathbb{R} \times (0,T)$. Since $uf_x = uf'(u)u_x$ is a perfect x derivative, and $u^{(\epsilon)}$ tends to zero as $|x| \to \infty$, we get, after integrating byparts that

(3.21) $\frac{1}{2} \int |u^{(\epsilon)}|^2 dx \Big|_0^T = -\epsilon \int_0^T |u^{(\epsilon)}_x|^2 dxdt$

It follows that

(3.21)' $\epsilon \|u^{(\epsilon)}_x\|^2 \equiv \epsilon \int_0^T \int |u^{(\epsilon)}_x|^2 dxdt \leqslant \text{const} = \frac{1}{2} \int u_0^2 dx.$

Clearly, $\epsilon\|u^{(\epsilon)}_x\| \leqslant \epsilon^{1/2}$ const tends to zero as $\epsilon\to 0$; this shows that div $h^{(\epsilon)}$ belongs to a compact set in $H^{-1}_{loc}$.

To construct $k^{(\epsilon)}$ we take any $C^2$ function $\eta(v)$, and define $\varphi(v)$ by

$$\varphi(v) = \int_{v_0}^{v} \eta'(u)f'(u)\lambda u.$$

Then

(3.22) $\varphi' = \eta' f'$

Multiply (2.17) by $\eta'$; using (3.22) we can write the resulting equation as

(3.23) $\eta^{(\epsilon)}_t + \varphi^{(\epsilon)}_x = \epsilon\eta' u^{(\epsilon)}_{xx} = \epsilon\eta^{(\epsilon)}_{xx} - \epsilon\eta'' u^{(\epsilon)2}_x$

We take now the vector function $k^{(\epsilon)}$ to be

(3.24) $k^{(\epsilon)} = \left[\varphi^{(\epsilon)}, -\eta^{(\epsilon)}\right]$

Using (3.23) we see that

(3.25) curl $k^{(\epsilon)} = \eta^{(\epsilon)}_t + \varphi^{(\epsilon)}_x = \epsilon\eta^{(\epsilon)}_{xx} - \epsilon\eta'' u^{(\epsilon)2}_x$.

We claim that the right side lies in a compact set in $H^{-1}_{loc}$. Clearly, the $H^{-1}_{loc}$ norm of the first term $\epsilon\eta^{(\epsilon)}_{xx}$ is bounded by

$$\epsilon \|\eta^{(\epsilon)}_x\| = \epsilon \|\eta' u^{(\epsilon)}_x\|$$

It follows, as before, from (3.21)' that this tends to zero. It follows from (3.21)' that the second term in (3.23), $\epsilon\eta'' u^{(\epsilon)2}_x$, is bounded in $L^1$. This does not imply $H^{-1}_{loc}$ compactness; to get that we note that since by Theorem 3.2, $u^{(\epsilon)}$ is uniformly bounded in $L^\infty$, it follows that so are $\eta^{(\epsilon)}$ and $\varphi^{(\epsilon)}$. It follows that curl $k^{(\epsilon)} = \eta_t^{(\epsilon)} + \varphi_x^{(\epsilon)}$ belong to a bounded set in $W^{-1,\infty}$. Now Tartar appeals to

Murat's Lemma:

Suppose the set of functions $\{c\}$ satisfies these conditions:

(a) $\{c\}$ belongs to a bounded set in $W^{-1,\infty}$.

(b) Each c can be decomposed as

$$c = c_1 + c_2$$

where $\{c_1\}$ belongs to a compact set in $H^{-1}_{loc}$, and $\{c_2\}$ to a bounded set in $L^1$.

Assertion: The set $\{c\}$ belongs to a compact subset of $H^{-1}_{loc}$.

Having verified the hypotheses of compensated compactness, we can conclude as in (3.18) that

(3.26) $u^{(\epsilon)}\varphi^{(\epsilon)} - f^{(\epsilon)}\eta^{(\epsilon)} \rightharpoonup u\varphi(u) - f(u)\eta(u)$

We use now formula (3.15) describing the weak limits of functions of $u^{(\epsilon)}$; taking $g(v) = v\varphi(v) - f(v)\eta(v)$ we see that the weak limit of the left side of (3.26) is

$$\langle v\varphi - f\eta, \nu\rangle$$

The various terms on the right of (3.26) can be similarly described; since the two sides of (3.26) are equal, we conclude that

(3.27) $\langle v\varphi - f\eta,\nu\rangle = \langle v,\nu\rangle\langle\varphi,\nu\rangle - \langle f,\nu\rangle\langle\eta,\nu\rangle,$

where $\nu = \nu(x,t)$; (3.27) holds for a.a.(x,t).

We introduce the abbreviations

$$(3.28) \quad u = \langle v, \nu \rangle, \qquad \bar{f} = \langle f(v), \nu \rangle;$$

then (3.27) can be rewritten as

$$(3.29) \quad \langle (v-u)\varphi, \nu \rangle = \langle (f-\bar{f})\eta, \nu \rangle.$$

In the derivation of (3.27) we used second derivatives of $\eta$, but (3.27) itself depends continuously on $\eta$ in the C topology; therefore, (3.37) remains true for $\eta$ piecewise $C^1$. We choose

$$(3.30) \quad \eta(v) = |v-u|;$$

then from (3.22)

$$(3.30)' \quad \varphi(v) = \begin{cases} f(v) - f(u) & \text{for } u>v \\ f(u) - f(v) & \text{for } u<v \end{cases}$$

Setting these choices for $\eta$ and $\varphi$ into (3.29) gives

$$\langle |v-u|(f-f(u)), \nu \rangle = \langle (f-\bar{f})\, |v-u, \nu \rangle;$$

we deduce from this that

$$(3.31) \quad (f(u)-\bar{f}) \langle |v-u|, \nu \rangle = 0.$$

For f strictly convex, it follows from (3.28) and Jensen's inequality that the first factor $(f(u)-\bar{f})$ in (3.31) is positive unless $\nu$ is concentrated at the single point u; the same, of course, is true for the second factor $\langle |v-u|, \nu \rangle$. Thus, it follows from (3.31) that $\nu$ is concentrated at the single point u, and so $u^{(\epsilon)}$ tends to u in the $L^1$

sense. It follows then, as shown at the beginning of this section, that $u = \lim u^{(\epsilon)}$ satisfies (2.1) in the sense of distributions.

The argument outlined above shows that for every sequence of $\epsilon \to 0$ we can select a subsequence such that the solutions $u^{(\epsilon)}$ of (2.17) with prescribed initial value $u_0$ tend in the $L^1$ sense to a distribution solution u of (2.1) with initial value $u_0$. To prove that $\lim_{\epsilon \to 0} u^{(\epsilon)}$ exist, we have to show that any two subsequences have the same limit. For this we need the following characterization of such limits, see [11]:

<u>Theorem 3.4</u>: Let u be the $L^1$ limit of a subsequence $u^{(\epsilon)}$ of solutions of (2.27). Let $\eta$ be any <u>convex</u> function, and $\varphi$ related to $\eta$ by (3.22). Then

$$\eta(u)_t + \varphi(u)_x \leqslant 0 \tag{3.32}$$

in the sense of distribution.

The proof follows from (3.23); for when $\eta$ is convex, $\eta'' \geqslant 0$, and so (3.23) implies

$$\eta_t^{(\epsilon)} + \varphi_x^{(\epsilon)} \leqslant \epsilon \eta_{xx}^{(\epsilon)} .$$

(3.32) is the limit in the distribution sense of this relation as $\epsilon \to 0$.

Condition (3.32) is called an <u>entropy condition</u>; this notion will be elaborated in Sections 5 and 6.

It can be shown that distribution solutions of (2.1) that satisfy the entropy conditions (3.32) are <u>uniquely determined</u> by their initial data. This proves that the L' limits of two different subsequences of $u^{(\epsilon)}$ are the same; this completes the proof of convergence.

Tartar observed that the argument above for the $L^1$ convergence of $u^{(\epsilon)}$ also proves the compact dependence of solutions u on their initial data. For let $u_o^{(n)}$ be a uniformly bounded sequence of data with common support, converging in the weak$^*$ topology of $L^\infty$ to $u_0$; denote by $u^{(n)}$ the corresponding solutions of (2.1) constructed by the viscosity method. A subsequence can be selected so that for every continuous g, $g(u^{(n)})$ converges in the weak$^*$ topology of $L^\infty$. Again, we look at the two vector valued functions

$$h^{(n)} = (u^{(n)}, f^{(n)}) \qquad \text{and} \qquad k^{(n)} = (\varphi^{(n)}, -\eta^{(n)})$$

By (2.1),

$$\text{grad } h^{(n)} = 0,$$

and by taking the limit of (3.5)

$$\text{curl } k^{(n)}$$

are measures, uniformly bounded. Using compensated compactness we conclude as before that $u^{(n)}$ converges in the $L^1$ norm to a limit u, a solution of (2.1) with initial value $u_0$. Since each $u^{(n)}$ satisfies the entropy inequality (3.32), taking the distribution limit of (3.32) shows that so does the limit u. Then by the uniqueness theorem quoted above we conclude that u is the solution of (2.1) with initial value $u_0$ obtained by the viscosity method. Thus, the $L^1$ limits of two different subsequences of $u^{(n)}$ are the same; this completes the proof of the compactness of the mapping $u_0 \to u$ from $L^\infty$ to $L^1$.

We close this section by remarking that compactness is a property one usually associates with the dependence of solutions of parabolic equations such as (2.17) on their initial data; unlike in the linear case, this property is preserved as $\epsilon$ tends to zero. It is instructive to see in some detail how this happens. From (3.21) we get that

$$(3.33) \quad \frac{d}{dt} \int \frac{|u^{(\epsilon)}|^2}{2} dx = -\epsilon \int |u^{(\epsilon)}|^2 dx,$$

i.e., that $\epsilon \int u_x^2 dx$ is the rate of energy dissipation. We saw earlier that the limit u of $u^{(\epsilon)}$ has discontinuities; it can be shown that, for $\epsilon$ small, $u^{(\epsilon)}$ bridges this discontinuity in a layer $\ell(\epsilon)$ of width $\epsilon$; in this layer it has the shape of a traveling wave:

$$u^{(\epsilon)}(x,t) \simeq w\left(\frac{x-st}{\epsilon},t\right), \tag{3.34}$$

where $w(\xi)$ is a solution of

$$-s\dot{w} + f'\dot{w} = \ddot{w}, \qquad \dot{} = \frac{d}{d\xi}.$$

Differentiating (3.34) we see that

$$\epsilon \int_{\ell(\epsilon)} |u_x^{(\epsilon)}|^2 dx \simeq \frac{1}{\epsilon}\int \dot{w}\left[\frac{x-st}{\epsilon}\right]^2 dx = \int \dot{w}(\xi)^2 d\xi.$$

This shows that the rate of energy dissipation does not tend to zero as $\epsilon$ tends to zero. This is in sharp contrast to the linear case:

$$u_t^{(\epsilon)} = \epsilon u_{xx}^{(\epsilon)},$$

when $u^{(\epsilon)}$ tends to a discontinuous limit. Here the transition layer has width $\epsilon^{1/2}$, and the shape of the wave is

$$u^{(\epsilon)}(x,t) \simeq w\left(\frac{x}{\epsilon^{1/2}}, t\right).$$

So the rate of energy dissipation

$$\epsilon \int_{\ell} |u_x^{(\epsilon)}|^2 \simeq \int \dot{w}\left[\frac{x}{\epsilon^{1/2}}\right]^2 dx = \epsilon^{1/2} \int \dot{w}^2 d\xi$$

tends to zero as $\epsilon$ tends to zero.

## 4. Consequences of the Shock Condition

In this section we study those distribution solutions of (2.1) all of whose discontinuities are shocks, i.e., satisfy condition (2.23). We present a result of James Glimm that shows that such solutions form a compact set.

The shock condition relates the signal velocities $a_L$ and $a_R$ to the left and right of the discontinuity with the velocity s with which the discontinuity propagates:

$$a_L > s > a_R. \tag{4.1}$$

(4.1) expresses the fact that characteristics, i.e., curves propagating with signal speed drawn in the forward direction, intersect the shock. It follows from this that characteristics drawn in the backward direction cannot intersect any shock. It follows from this that in a solution all whose discontinuities are shocks, every point (x,t) can be connected by a characteristic to a point on the initial line.

For simplicity we shall study solutions u(x,t) whose initial data--and, therefore, themselves--are periodic functions of x with period L:

$$u(x + L,t) = u(x,t).$$

We shall estimate the total variations of u(x,t) with respect to x per period. We start by estimating the total variation of $a = a(u(x,t))$; the total variation is the sum of the increasing variation $A^+$ and the decreasing variation $A^-$; for a period function the two are the same:

$$\text{Total variation of } a \text{ per period at time } t = 2A^+(t) \tag{4.2}$$

To estimate the increasing variation of a we note that according to the shock condition (4.1), the discontinuities of a contribute only to the decreasing variation of a. Therefore, we can calculate $A^+(t)$ by dividing an interval of length L at t into subintervals by points

$$x_0(t) < x_1(t) < \cdot < x_N(t) = x_0(t) + L$$

so that a is alternately increasing and decreasing along each interval. Then

(4.3) $\quad A^+(t) = \Sigma\ a_{2n+1} - a_{2n},$

where

(4.4) $\quad a_j = a(u(x_j,t)).$

We connect now the points $x_j,t$ to points $y_j$ on the initial line by characteristics:

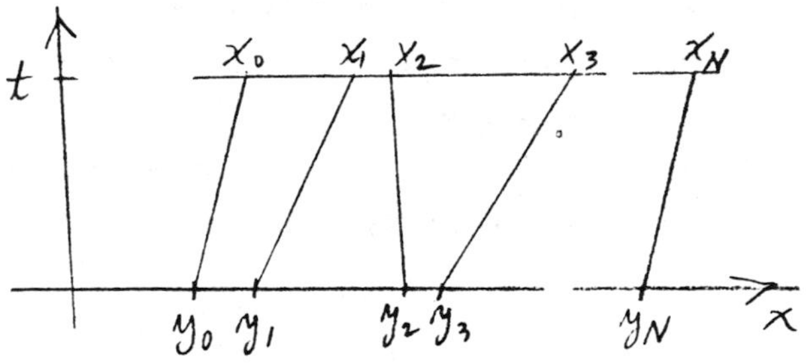


By (2.5)

(4.5) $\quad x_j(t) = y_j + a_j t$

Since $x_N = x_0 + L$,

$$\Sigma\ x_{2n+1}^{(t)} - x_{2n}(t) \leqslant L$$

Using formula (4.5) for $x_j$ we get

$$\Sigma\ y_{2n+1} - y_{2n} + t\ \Sigma\ a_{2n+1} - a_{2n} \leqslant L$$

Since the first sum is positive, we conclude, using formula (4.3), that

$$A^+(t) \leqslant \frac{L}{t}.$$

So by (4.2) we conclude that

$$\text{(4.6)} \quad \text{Total variation of } a(u(x,t)) \text{ per period} \leqslant \frac{2L}{t}$$

We use now the fact that (2.1) is genuinely nonlinear, i.e., that a is nonconstant function of u. Denote R a lower bound for $|a'(u)|$; then Total Var $a(u) \geqslant R$ Total Var u, so we conclude from (4.6):

<u>Theorem</u> <u>4.1</u> <u>(Glimm)</u>: Suppose equation (2.1) is genuinely nonlinear in the sense that

$$\text{(4.7)} \quad |f''(u)| \geqslant R.$$

Let u be a distribution solution of (2.1) all whose discontinuities are shocks, and which is periodic in x with period L. Then the total variation of u(x,t) in x per period is

$$\text{(4.8)} \quad \leqslant \frac{2L}{Rt}.$$

It follows from (4.8) that for any $t>0$, the measures $u_x(x,t)$ are uniformly bounded and, therefore, weakly compact; it is remarkable that nothing need be assumed about the initial data.

We can obtain a compactness result about the u themselves if we can get bound for their integral. This is easy because of the conservation form of the equation: if we integrate (2.1) with respect to x, we deduce that

$$\text{(4.9)} \quad \int_0^L u(x,t)dx$$

is independent of t.

There is an analogue of (4.8) for solutions that are not periodic in x but of compact support. The total variation of u(x,t) in x is less than

$$(4.8)' \quad \frac{C}{Rt^{1/2}},$$

where the constant depends on the length of the support of the initial data but on nothing else, see [12].

We conclude this section by showing that solutions which satisfy the shock condition (4.1) satisfy the entropy condition (3.32) as well, and conversely. Suppose that f is convex, so that a is an increasing function of u; then (4.1) is equivalent with

$$(4.10) \quad u_L > u_R.$$

For a piecewise continuous solution, the left side of (3.32) is zero except along a discontinuity $x = y(t)$, where it has the value

$$(4.11) \quad (\eta_L - \eta_R)s - (\varphi_L - \varphi_R)$$

times $\delta(x-y(t))$; here s denotes the shock velocity

$$s = \frac{dy}{dt}.$$

Since by (4.10), $u_R < u_L$, we can write (4.11) as

$$(4.11)' \quad \int_{u_R}^{u_L} (\eta' s - \varphi')dw.$$

Using relation (3.22) this can be written as

$$(4.11)'' \quad \int_{u_R}^{u_L} \eta'(s - f')du$$

We recall now the Rankine-Hugoniot condition (2.9):

$$s(u_L - u_R) = f_L - f_R;$$

this can be expressed by saying that

$$f(u) - su$$

has the same value at $u_L$ and $u_R$; denoting this common value by g we can integrate (4.11)" by parts to obtain

$$\text{(4.11)'''} \qquad \int_{u_R}^{u_L} \eta'' \; (f - su - g)dn$$

Since $\eta$ is assumed convex, $\eta'' > 0$; since f is assumed convex, it lies below any secant, i.e.

$$f < su + g$$

for u between $u_R$ and $u_L$. This shows that (4.11)''' is < 0; thus, so is (4.11), and the entropy condition (3.32) is fulfilled. The proof of the converse is the same.

## 5. Systems of Conservation Laws

In this section we note very briefly the extension of the ideas in Sections 2, 3 and 4 to systems discussed in Section 1. First of all, we have to assume that the matrix A given by (1.3) and appearing in Equation (1.2),

$$\text{(5.1)} \qquad u_t + Au_x = 0,$$

has real and distinct eigenvalues $\alpha_1 < \alpha_2 < \ldots < \alpha_n$; this makes equation (5.1) hyperbolic. The eigenvalues $\alpha$ are functions of u, and so are the eigenvectors r:

(5.2) $Ar = \alpha r$.

Genuine nonlinearity, see [8], is defined to be

(5.3) $\alpha_j' \cdot r_j \neq 0$

for $j = 1, 2, \ldots, n$; here ' denotes the gradient with respect to u.

The curves that satisfy

(5.4) $$\frac{dx}{dt} = \alpha_j$$

are still called characteristics, but they are no longer straight lines in general.

A function w of u is caled a <u>Riemann</u> <u>invariant</u> for the $j^{th}$ field if i for any solution u of (5.1), w(u) is constant along characteristics of the $j^{th}$ field. It follows easily that this is the case if w', the gradient of w, is a left eigenvector of A:

(5.5) $w'A = \alpha_j w'$.

For then multiplying (5.1) by w' gives

(5.5)' $w_t + \alpha_j w_x = 0$.

Since eigenvectors can be rescaled, system (5.5) can always be solved when $n = 2$; for $n>2$ solutions exist only in exceptional cases.

As in the scalar case, solutions in the classical sense cannot be continued beyond a finite time, see [10] and [6], so again we have to turn to distribution solutions. As before, a piecewise continuous solution is a distribution solution if the Rankine-Hugoniot condition (2.9) is satisfied for all n components. Again, the class of distribution solutions is too broad, and has to be narrowed by imposing some criterion of acceptance and rejection. The ones that make physical and mathematical sense are the same ones that were employed for scalar equations:

(a) The acceptable solutions are limits of solutions of a family of equations obtained by augmenting the flux by a small viscous term, and letting the viscosity tend to zero. The simplest way of augmenting the flux leads to the parabolic systems

(5.6) $u_t + f_x = \epsilon u_{xx}$,

or more generally

(5.6)' $u_t + f_x = \epsilon D u_{xx}$,

D some symmetric, positive matrix. We remark that physical viscosity and heat conductivity lead to a matrix D that is merely nonnegative, and is a function of u. For this reason equation (5.6) is said to contain <u>artificial</u> <u>viscosity</u>.

In [11], an entropy condition is formulated for limits of solutions of (5.6). Two functions $\eta(u)$ and $\varphi(u)$ are needed, with these properties:

(i) $\eta$ and $\varphi$ satisfy the differential equation

(5.7) $\eta' A = \varphi'$,

where $\eta'$ and $\varphi'$ denoted gradients with respect to u.

(ii) $\eta(u)$ is strictly convex.

It follows from (i) that if $u^{(\epsilon)}$ satisfies (5.6), $\eta^{(\epsilon)} = \eta(u^{\epsilon})$ and $\varphi^{(\epsilon)} = \varphi(u^{(\epsilon)})$ satisfy the vector analogue of (3.23). From this relation and (ii) we deduce that the $L^1$ limit u of $u^{(\epsilon)}$ satisfies the entropy inequality (3.32).

For n = 2, equation (5.7) has many convex solutions. For n > 2, (5.7) is an overdetermined system which only exceptionally has a solution; these exceptional cases happily include most of physical interest, see Section 6.

Theorem 3.1 has no analogy for systems; nor does E. Hopf's trick of linearizing (5.6) work for $n > 1$. Happily Tartar's approach is not tied to $n = 1$, and in fact Ron Di Perna succeeded in extending it to significant cases for $n = 2$, see [1] and [2].

(b) The finite difference approach analogous to (2.19) can be set up, as well as other more general ones. These work very well in practice--some better than others--but it is very hard to prove anything rigorously about their convergence. In [5] Glimm succeeded in proving the a.e. convergence of a scheme that is the mixture of a discrete scheme due to Godunov and a Monte Carlo-type scheme.

(c) The notion of a shock can be extended to piecewise continuous solutions of hyperbolic systems of conservation laws, as follows. In place of condition (4.1) we require that there be an index k such that

$$\begin{aligned} &a_k (u_L) > s > a_{k-1} (u_L), \\ &a_{k+1} (u_R) > s > a_k (u_R) \end{aligned} \tag{5.8}$$

Thus, there are n distinct families of shocks, corresponding to $k = 1, \ldots, n$.

In a k-shock, $n - k + 1$ characteristics impinge from the left on the line of discontinuity, and k from the right, altogether $n + 1$. The information carried by these characteristics combined with the Rankine-Hugoniot relations (2.9) serve to determine uniquely the solution on either side of the curve of discontinuity, and determine the curve itself.

Glimm's method for estimating the total variation of solutions sadisfying the shock condition, described in Section 4, has been extended in [5] to solutions with small oscillation of pairs of conservation laws.

It was shown in [11] that a piecewise continuous solution that satisfies the shock condition also satisfies the entropy condition, at least for sufficiently weak shocks.

## 6. Thermodynamics and Gas Dynamics

There are many thermodynamical variables; some of them, like

Density, $\rho$
Temperature, T
Pressure, p
Soundspeed, c

are quantities palpable in everyday experience. Others, such as

Internal Energy, e
Entropy, S
Enthalpy, i

per unit mass, are theoretical constructs. In a single-phase system at thermodynamic equilibrium any three are functionally related, i.e., are linked by an equation called equation of state. Any of the variables can be expressed as function of any other two.

The equations governing the flow of compressible gas are the laws of conservation of mass momentum and energy. Let $\rho$, m and E denote mass, momentum and total energy per unit volume, u the velocity of the flow; set

$$(6.1) \qquad m = \rho u, \ E = \rho e + \frac{1}{2} \rho u^2.$$

The conservation laws for flows depending on a single space variable x are:

$$(6.2) \qquad \begin{aligned} &\rho_t + m_x = 0 \\ &m_t + (um + p)_x = 0 \\ &E_t + (u(E + p))_x = 0. \end{aligned}$$

It is not hard to show that this system is hyperbolic; the three signal velocities are

$$(6.3)\qquad a_1 = u - c,\ a_2 = u,\ a_3 = u + c,\ c = \left(\frac{\partial p}{\partial \rho}\right)^{1/2}$$

$a_1$ and $a_3$ are genuinely nonlinear in the sense of (5.3). The signal velocity $a_2 = u$ is linearly degenerate; the corresponding characteristic curves are particle paths.

It follows that if equations (6.2) are satisfied at every point, then

$$(6.4)\qquad S_t + uS_x = 0.$$

In words: in a smooth flow entropy per unit mass is constant along particle paths.

Combining equation (6.4) with $(6.2)_1$ we deduce the conservation law

$$(6.5)\qquad (\rho S)_t + (u\rho S)_x = 0$$

for entropy per unit volume. It is not hard to show that $\rho S$ is a <u>concave</u> function of $\rho$, m and E; since (6.5) is a consequence of (6.1), the pair

$$\eta = \rho S \text{ and } \varphi = u\rho S$$

must satisfy relation (5.7). Thus, it follows that distribution solutions of (6.1) which are the limits of artificially viscous flows (5.6) satisfy the entropy inequality (3.32):

$$(6.6)\qquad (\rho S)_t + (u\rho S)_x \geqslant 0.$$

We have seen in equation (4.11) that for a piecewise continuous solution relation (6.6) means that across each discontinuity

$$[(\rho S)_L - (\rho S)_R]\, s - (u\rho S)_L + (u\rho S)_R \geqslant 0$$

Rearranging turns this into

$$(6.7) \quad (s - u_L)\rho_L S_L - (s - u_R)\rho_R S_R \geqslant 0.$$

According to the Rankine–Hugoniot relation applied to the conservation of mass equation $(6.1)_1$,

$$s(\rho_L - \rho_R) = (u\rho)_L - (u\rho)_R,$$

which can be rearranged as below:

$$(6.8) \quad (s - u_L)\rho_L = (s - u_R)\rho_R.$$

We turn now to the shock relations (5.8); for $k = 1$ we get from (5.8), using (6.3), that

$$(6.9)_1 \quad u_L > s, \qquad u_R > s$$

while for $k = 3$ we get that

$$(6.9)_3 \quad u_L < s, \qquad u_R < s$$

Thus in case $k = 1$, (6.8) is negative, in case $k = 3$, (6.8) is positive. Setting this into (6.7) we conclude that for $k = 1$

$$(6.10)_1 \quad S_L < S_R,$$

while for $k = 3$,

$$(6.10)_3 \quad S_L > S_R.$$

It follows from $(6.9)_1$ that in case $k = 1$, particles cross the shock from left to right, while $(6.9)_3$ says that in case $k = 3$ particles cross the shock from right to left. So both $(6.10)_1$ and $(6.10)_3$ can be summarized in the single statement

Theorem 6.1: When gas crosses a shock, its entropy increases.

We shall now derive an integral form of the increase of entropy. We assume that the gas is in the same state and that it travels at the same velocity at $x = \infty$ as at $x = -\infty$. We normalize entropy so that its value = 0 at $x = \pm\infty$. Then integrating (6.6) with respect to x gives

$$\frac{d}{dt} \int \rho S \, dx \geqslant 0; \tag{6.11}$$

in words.

Theorem 6.2: The total amount of entropy in the flow field is an increasing function of time.

We compare now this result with Glimm's result in Section 4, according to which the total variation of a solution tends to decrease with time. This decrease of total variation can be thought of as a loss of detail, or loss of information. Ever since Maxwell's demon, Boltzmann's H theorem, Smoluchowsky's study of fluctuations, Szilard's thesis and Shannon's work on information and its transmission there have been many attempts to link increase of entropy to decrease of information about the gas. We have shown here that shock waves do both: increase entropy and decrease information, although not on the molecular level but pertaining to organized motion in the whole flow-field.

We turn now to another manifestation of the second law of thermodynamics,

Carnot's Theorem: Consider an engine that extracts heat energy in the amount $Q_h$ from a hot reservoir whose temperature is $T_h$, and dumps $Q_c$ amount of heat energy into a cold reservoir whose temperature is $T_c$; the difference is turned into mechanical energy

$$W = Q_h - Q_c.$$

The engine operates cyclically, i.e., everything is in the same state at the end of the cycle as at the beginning.

Assertion: The efficiency of such an engine in converting heat energy into mechanical energy cannot exceed

$$\frac{T_h - T_c}{T_h}$$

Proof: During the absorption of heat energy in the amount of $Q_h$ from a reservoir at temperature $T_c$, the engine's entropy is increased by the amount

$$(6.12)\quad S = \frac{Q_h}{T_h}$$

Since at the end of the cycle the engine returns to its original state, this amount of entropy must be returned to the outside when $Q_c$ amount of heat is released to a reservoir at temperature $T_c$. Assuming that no further entropy is generated during the cycle,

$$(6.12)'\quad S = \frac{Q_c}{T_c}.$$

The amount of mechanical energy W extracted is the difference of the heat energy absorbed and released:

$$W = Q_h - Q_c.$$

Using (6.12) and (6.12)' we get the following formula for the efficiency $W/Q_h$:

$$(6.13)\quad \frac{W}{Q_h} = \frac{Q_h - Q_c}{Q_h} = \frac{ST_h - ST_c}{ST_h} = \frac{T_h - T_c}{T_h}$$

This proves the theorem. Note that if there is entropy production during the cycle, e.g., by shocks produced by the operation of the engine, then an additional amount S' of entropy is produced and has to be gotten rid of. In this case, relation (6.12)' has to be modified to

$$(6.12)'' \quad S + S' = \frac{Q_c}{T_c}.$$

Clearly, this lowers the efficiency to

$$\frac{W}{Q_h} = \frac{T_h - T_c\ (1+S'/S)}{T_h}$$

Clearly, for sake of efficiency one must keep the entropy production and, thus, the strength of shock waves formed during the operation of the engine to a minimum.

What does Carnot's theorem imply for gas dynamics? Imagine an infinite tube with unit cross section, filled from a cold reservoir, except for a finite section (-L,L) which is filled from a hot reservoir. The initial data are

$$u_0(x) \equiv 0$$

$$(6.14) \quad e_0(x) = \begin{cases} e_h & \text{for } |x|<L \\ e_c & \text{for } |x|>L \end{cases}$$

$$T_0(x) = \begin{cases} T_h & \text{for } |x|<L \\ T_c & \text{for } |x|>L \end{cases}$$

The added heat energy is

$$2L\rho(e_h - e_c).$$

According to Carnot's theorem, the amount that may be converted into mechanical energy does not exceed

$$2L\rho_h e_h \frac{T_h - T_c}{T_h}.$$

Therefore, kinetic energy, which is mechanical energy, can at no future time exceed this amount.

This argument can be extended to variable initial data:

$$u_0(x) = 0 \text{ for } |x|>L$$

$$(6.14)' \quad e_0(x) \begin{cases} \geq e_c & \text{for } |x|< \\ = e_c & \text{for } |x|> \end{cases}$$

$$T_0(x) \begin{cases} \geq T_c & \text{for } |x|<L \\ = T_c & \text{for } |>L. \end{cases}$$

We can think of constructing such data by filling a finite section of the tube with gas taken from a collection of reservoirs and setting it in motion. The amount of heat energy that can be converted to mechanical energy is

$$(6.15) \quad \int \rho_0(x) e_0(x) \frac{T_0(x) - T_c}{T_0(x)} dx.$$

Kinetic energy at any time $t>0$ cannot exceed this amount plus the kinetic energy initially imparted to the system:

<u>Theorem</u> <u>6.3</u>: Let u, $\rho$, e, T denote the velocity, density, internal energy and temperature of a gas moving in a tube, whose initial data satisfy the restrictions (6.14)'. Then for all $t>0$

$$(6.16) \quad \frac{1}{2} \int \rho u^2 dx \leq \frac{1}{2} \int \rho_0 u_0^2 dx + \int \rho_0 e_0 \frac{T_0 - T_c}{T_0} dx.$$

This inequality holds in any number of dimensions; it seems desirable to obtain a proof of it by PDE methods.

We conclude by recounting a curious result about strong shocks that seem to echo a theme in kinetic theory. The Rankine–Hugoniot relations (2.9) for the conservation laws of momentum and energy, see (6.1), are

$$
\begin{aligned}
&s[m] = [um+p], \\
(6.17) \qquad & \\
&s[E] = [u(E+p)].
\end{aligned}
$$

Assume that a <u>strong shock</u> is impinging on a gas <u>at rest</u>; we denote the quantities behind the shock by capital letters: E, M, P, in the front by lower case letters: e, m, p; note that $u = m = 0$. We rewrite (6.17) as

$$sM = UM+P-p$$

$$s(E-e) = UE+UP$$

We solve the first equation for s and set it into the second:

$$\left(U+\frac{P-p}{M}\right)(E-e) = UE+UP,$$

from which

$$\frac{P-p}{M}(E-e) = U(P+e)$$

Multiplying by $\frac{M}{P}$ we get

$$(6.18) \qquad \left(1-\frac{p}{P}\right)\left(1-\frac{e}{E}\right) E = MU\left(1+\frac{e}{P}\right)$$

For a strong shock, $p/P$, $e/E$ and $e/P$ are small; so we get

$$(6.18)' \qquad E = MU\,(1+\epsilon),$$

$\epsilon$ small. Thus, the total energy E behind the shock is approximately twice the kinetic energy $\frac{1}{2}MU$. In words:

Theorem 6.4: When a strong shock impinges on a gas at rest, the energy imparted to the gas is equipartitioned; that is, approximately half of it goes into kinetic, the other half into internal energy.

Note that it follows from (6.18), (6.18)' that the internal energy is always a little greater than the kinetic.

One would like to know what this kind of equipartition of energy has to do with the equipartitioning that is the hallmark of thermodynamic equilibrium.

## Bibliography


[1] DiPerna, R. J., Convergence of approximate solutions to conservation laws. Arch. Rat. Mech. Anal., Vol 82, 1983, pp 27-70.

[2] DiPerna, R. J., Convergence of the viscosity method for isentropic gas dynamics, Comm. Math. Phys., 1983, Vol 91, pp 1-30.

[3] Douglis, A., Layering methods for nonlinear partial differential equations of first order, Ann. Inst., Fourier Grenoble, Vol 22, 1972, pp 141-227.

[4] Glimm, J., Solutions in the large for nonlinear hyperbolic systems of equations, Comm. Pure Appl. Math. 18, 1965, pp 697-715.

[5] Glimm, J., Lax, P. D., Decay of solutions of systems of nonlinear hyperbolic conservation laws, Mem. Am. Math. Soc. 101, 1970.

[6] John, F., Formation of singularities in one-dimensional nonlinear wave propagation, Comm. Pure Appl. Math., Vol 27, 1974, pp 377-405.

[7] Keyfitz, B., Solutions with shocks: An example of an L'-contractive semigroup, Comm. Pure Appl. Math., Vol 24, 1971, pp 125-132.

[8] Lax, P. D., Weak solutions of nonlinear hyperbolic equations and their numerical computation, Comm. Pure Appl. Math. 7, 1954, pp 159-193.

[9] Lax, P. D., Hyperbolic systems of conservation laws, II, Comm. Pure Appl. Math. 10, 1957, pp 537-566.

[10] Lax, P. D., Development of singularities of solutions of nonlinear hyperbolic partial differential equations, J. Math. Phys., Vol 5, 1964, pp 611-613.

[11] Lax, P. D., Shock waves and entropy, in Contributions to Nonlinear Functional Analysis, ed. E. A. Zarantonello, Academic Press, 1971, pp 603-634.

[12] Lax, P. D., The formation and decay of shock waves, Amer. Math. Monthly, Vol 79, 1972, pp 227-241.

[13] Lax, P. D., Hyperbolic systems of conservation laws, and the mathematical theory of shock waves, Regional Conf. Series in Appl. Math., 1973, SIAM, Phila.

[14] Lax, P. D., Accuracy and resolution in the computation of solutions of linear and nonlinear equations, Recent Adv. in Num. Anal., Acad. Press, 1978, pp 107-117.

[15] Murat, F., Compacité par compensation, Ann. Scuola Norm. Sup. Pisa Sci. Fis. Mat. 5, 1978, pp 489-507.

[16] Smoller, J., Shock waves and reaction-diffusion equation, 1982, Springer Verlag.

[17] Tartar, L., Compensated compactness and applications to partial differential equations, in Research Notes in Mathematics 39, Nonlinear analysis and mechanics: Heriot-Watt Symposium, Vol 4, ed. R. J. Knops, Pitman Press, 1979.

# Stress and Riemannian Metrics in Nonlinear Elasticity

By

Jerrold E. Marsden*

Department of Mathematics

University of California

Berkeley, California 94720

## §1. Introduction

In Doyle and Ericksen [1956, p. 77] it is observed that the Cauchy stress tensor $\sigma$ can be derived by varying the internal energy e with respect to the Riemannian metric on space: $\sigma^{ab} = 2\rho\partial e/\partial g_{ab}$. Their formula has gone virtually unnoticed in the elasticity literature. In this lecture we shall explain some of the reasons why this formula is, in fact, of fundamental significance. Some additional reasons for its importance follow. First of all, it allows for a rational derivation of the Duhamel-Neumann hypothesis on a decomposition of the rate of deformation tensor (see Sokolnikoff [1956, p. 359]), which is useful in the identification problem for constitutive functions. This derivation, due to Hughes, Marsden and Pister, is described in Marsden and Hughes [1983, p. 204-207]. Second, it is used in extending the Noll-Green-Naghdi-Rivlin balance of energy principle (using invariance under rigid body motions) to a covariant theory which allows arbitrary mappings. This is described in Section 2.4 of Marsden and Hughes [1983] and is closely related to the discussion herein. Finally, in classical relativistic field theory, it has been standard since the pioneering work of Belinfante [1939] and Rosenfeld [1940] to regard the stress-energy-momentum tensor as the derivative of the Lagrangian density with respect to the spacetime (Lorentz) metric; see for example, Hawking and Ellis [1973, Sect. 3.3] and Misner, Thorne and Wheeler [1973, Sect. 21.3]. This modern point of view has largely replaced the construction of "canonical

*Research partially supported by DOE Contract DE-AT03-82ER12097. Lecture given in Chern's Mathematical Sciences Research Institute PDE Seminar, April 4, 1983.

stress-energy-momentum tensors". Thus, for the Lagrangian formulation of elasticity (relativistic or not) the Doyle-Ericksen formulation plays the same role as the Belinfante-Rosenfeld formula and brings it into line with developments in other areas of classical field theory.

Acknowledgements. I thank John Ball, Tom Hughes, Morton Gurtin and Juan Simo for helpful comments.

## §2. Some Basic Notation

Let $\mathcal{B}$ and $\mathcal{S}$ be oriented smooth n-manifolds (usually n = 3). We call $\mathcal{B}$ the reference configuration and $\mathcal{S}$ the ambient space. Let $\mathcal{C}$ denote the set of smooth embeddings of $\mathcal{B}$ into $\mathcal{S}$, so $\phi \in \mathcal{C}$, $\phi:\mathcal{B} \to \mathcal{S}$ represents a possible configuration of an elastic body. (In many situations one needs to put Sobolev or Hölder differentiability conditions on elements of $\mathcal{C}$; such conditions will not interfere with, nor play a significant role in our discussions, so we work with $C^\infty$ objects for simplicity.)

A motion is a curve $\phi(t) \in \mathcal{C}$. For $X \in \mathcal{B}$, we write $x = \phi(X,t) = \phi(t)(X)$. The material velocity of a motion is the curve of vector functions over $\phi(t)$ defined by

$$V(t)(X) \equiv V(X,t) = \frac{\partial}{\partial t} \phi(X,t);$$

thus $V(t)(X) \in T_x\mathcal{S}$, the tangent space to $\mathcal{S}$ at x. The spatial velocity is $v(t) = V(t)\circ\phi(t)^{-1}$, a vector field on the image $\phi(t)(\mathcal{B}) \subset \mathcal{S}$.

Let $\mathcal{B}$ and $\mathcal{S}$ carry Riemannian metrics G and g, respectively, and associated volume elements dV and dv. Using the standard pull-back notation from analysis on manifolds (see Abraham, Marsden and Ratiu [1983]), we let $C = \phi^*g$, the Cauchy-Green deformation tensor. In coordinates $\{X^A\}$ on $\mathcal{B}$ and $\{x^a\}$ on $\mathcal{S}$, if $F^a{}_A$ denote the components of the derivative $F = T\phi$, then $C_{AB} = g_{ab}F^a{}_AF^b{}_B$ (for the relationship with the formula $C = F^TF$, see Marsden and Hughes [1983, Sect. 1.3]).

Let us recall the usual approach to the Cauchy equations of elasticity (see Truesdell and Noll [1965]). Let W be a materially frame indifferent stored energy function; that is, W is a function of X $\in \mathcal{B}$ and the point values of C. We write W(X,C) and let $\rho_{\mathrm{Ref}}$ denote the mass density in the reference configuration. Let

$$P = \rho_{\mathrm{Ref}} \frac{\partial W}{\partial F}$$

denote the <u>first Piola-Kirchhoff stress tensor</u> and A(t) be the material acceleration of a motion $\phi(t)$ (the acceleration is defined using the Levi-Civita connection of g). Cauchy's equations are:

$$\rho_{\mathrm{Ref}} A = \mathrm{DIV}\ P + B$$

where B is an external body force. In the spatial picture these equations read

$$\rho a = \mathrm{div}\ \sigma + b$$

where $\rho_{\mathrm{Ref}} = J\rho$ (J is the Jacobian of $\phi$), $J\sigma = PF^T$, so $\sigma$ is the Cauchy stress, and b is an external spatial body force. These equations are usually derived by postulating the integral form of balance of momentum and sufficient smoothness. (There are also boundary conditions to be imposed, but we shall not explicate them here.)

Although one can do it, it is not trivial or especially natural to pass directly from the integral balance of momentum assumption to the weak form of Cauchy's equations; see Antman and Osborne [1979]. Another difficulty with balance of momentum is that it does not make sense on manifolds; in particular, this approach is not useful for studying continuum mechanics in general relativity.

Let us now outline some different approaches based on covariant energy principles which overcome the preceding objections.

## §3. Energy Principles in the Material Picture

Let us first recall how one introduces stress by an energy principle. Let W be a given materially frame indifferent stored energy function. Let us regard W as a function of $\phi$, g and G. Assume that

1. W is spatially covariant; that is, for any diffeomorphism $\xi$ of $S$,

$$W(\phi,g,G) = W(\xi \circ \phi, \xi_* g, G)$$

and

2. W is local; that is, if $(\phi_1,g_1,G_1)$ and $(\phi_2,g_2,G_2)$ agree in a neighborhood of X (with g evaluated at $x = \phi(X)$), then $W(\phi_1,g_1,G_1)(X) = W(\phi_2,g_2,G_2)(X)$.

As in the usual Coleman-Noll [1959] argument these assumptions together with energy balance laws below enable one to deduce that W depends only on the point values of C. For simplicity let us assume this at the outset.

Notice that we do not necessarily assume that W is materially covariant; that is, $W(\phi \circ \eta, g, \eta^* G) = W(\phi,g,G) \circ \eta$ for diffeomorphisms $\eta$ of $B$. Indeed, this holds if and only if the material is isotropic.

The energy function is

$$\mathcal{H}(U,\phi,g,G) = \int_U \rho_{Ref}(\frac{1}{2} \|V(t)\|^2 + W(\phi,g,G))dV$$

where $U \subset B$ is a compact region with smooth boundary.

Now assume that there is a traction field, namely for each $\phi$,g,G we have a map $T(\phi,g,G):TB \to T^*S$ covering $\phi$ such that the following holds:

Cauchy's Axiom of Power. Admissible motions satisfy the following condition:

$$\frac{d}{dt}\mathcal{H}(U,\phi,g,G) = \int_{\partial U} \langle T(\phi,g,G)(N),V(t)\rangle dA + \int_U B\cdot V(t)dV$$

<u>where</u> N <u>is</u> <u>the</u> <u>outward</u> <u>unit</u> <u>normal</u> <u>to</u> $\partial U$.

Cauchy's theorem states that if this holds for all U, then necessarily T is linear in N and so defines a two tensor P.

A. <u>The Hamiltonian Systems Approach</u>

In this approach we first prove that $P = \rho_{Ref}\partial W/\partial F$ by assuming that there are enough motions so that V can be varied arbitrarily at any fixed $\phi$; for example one often assumes "any" motion is possible by choosing $B$ appropriately. Then the divergence theorem shows that indeed, $P = \rho_{Ref}\partial W/\partial F$. With this equation in hand one can then assume that motions come from a Hamiltonian system (adapted to take care of external forces) on $T\mathcal{C}$ or $T^*\mathcal{C}$ with energy function $\mathcal{H}(B,\phi,g,G)$. These yield directly the <u>weak</u> form of the Cauchy equations (cf. Marsden [1981, Lectures 1 and 2]).

B. <u>Covariance Approach</u>

In the covariant approach one assumes that an admissible motion $\phi(t)$ satisfies Cauchy's axiom of power and that for any curve $\xi(t)$ of diffeomorphisms of $S$, the new motion $\phi'(t) = \xi(t)\circ\phi(t)$ also satisfies balance of energy in which g is replaced by $\xi_* g$ and the velocities, forces and accelerations are transformed according to the standard dictates of the Cartan theory of classical spacetimes (see Marsden and Hughes [1983, Sect. 2.4]). In the material picture this approach directly yields a version of the weak form of the equations and a formula for the "rotated stress":

$$\Sigma = 2\rho\partial W/\partial G, \text{ i.e., } \Sigma^{AB} = 2\rho\frac{\partial W}{\partial G_{AB}},$$

which is a material version of the Doyle-Ericksen formula. See Simo and Marsden [1984] for details.

Remark. The symmetry of the stress is built into the assumption of material frame indifference. Indeed, if $S = F^{-1}P$ is the second Piola-Kirchhoff stress, then one finds from the chain rule that

$$S = 2\rho_{\text{Ref}}\frac{\partial W}{\partial C}$$

which is symmetric; see the appendix. This formula for the stress is related to the covariance approach in the convected picture; again see Simo and Marsden [1984].

## §4. Energy Principles in the Spatial Picture

Spatially, the energy is also regarded as a function of $\phi$, g and G by

$$e(\phi,g,G) = W(\phi,g,G)\circ\phi^{-1}$$

Balance of energy for a moving region $\phi(t)(U) = U(t)$ where $U \subset \mathcal{B}$, now takes the form

$$\frac{d}{dt}\int_{U(t)} \rho\left(e + \frac{1}{2}\|v\|^2\right)dv = \int_{\partial U(t)} t\cdot v\ da + \int_{U(t)} b\cdot v\ dv$$

The spatial form of Cauchy's axiom of power states that admissible motions satisfy the preceeding equation. As before, this implies that t, the Cauchy traction vector, is linear in n, the unit outward normal to $\partial U(t)$, so defines a two tensor $\sigma$ (depending on $\phi$,g and G), the Cauchy stress tensor.

To complete the spatial description two routes are possible.

### A. The Hamiltonian Systems Approach

If one assumes that any motion is possible with suitable forces, then as before, one gets an equation for the stress. Using the spatial picture, this equation is

$$\sigma = 2\rho\frac{\partial e}{\partial g}, \text{ i.e., } \sigma^{ab} = 2\rho\frac{\partial e}{\partial g_{ab}}$$

In this approach one then assumes that the variables $(m,\rho,C)$, where $m = \rho v$ is the momentum density, form a Hamiltonian system in a sense involving Lie-Poisson brackets for Lie groups analogous to the way the Euler equations for a rigid body are Hamiltonian when written in terms of its three angular velocities or momenta. See Holm and Kuperschmidt [1983] and Marsden, Ratiu and Weinstein [1984] for details. As before, this directly yields the weak form the spatial equations.

B. The Covariant Approach

In this approach we assume that a motion satisfies balance of energy and that balance of energy is still valid under any superposed curve of diffeomorphisms $\xi(t)$, where as above, the superposed motion uses the metric $\xi_* g$ and the other quantities are transformed by the dictates of classical mechanics. This assumption directly yields the weak form of the evolution equations, conservation of mass and the Doyle-Ericksen formula $\sigma = 2\rho\partial e/\partial g$. The proof of this is similar to that of Theorem 4.13 in Marsden and Hughes [1983].

§5. Concluding Remarks

1. Since all of the approaches sketched are equivalent, the four basic formulas for the stress

$$P = \rho_{Ref}\frac{\partial W}{\partial F}, \quad \sigma = 2\rho\,\frac{\partial e}{\partial g},$$

$$S = 2\rho\ Ref\ \frac{\partial W}{\partial C}, \text{ and } \Sigma = 2\rho\,\frac{\partial W}{\partial G}$$

must be equivalent as well. Indeed, their equivalence is easily checked directly using the chain rule and the relation $C_{AB} = g_{ab}F^a{}_AF^b{}_B$. (This chain rule argument is how Doyle and Ericksen [1956] present the formula.) This is detailed in the appendix for the first three formulas. (The fourth requires special interpretations and is omitted only for brevity.)

2. Some special peculiarities with the Hamiltonian formalism arise when one is considering electromagnetic fields coupled to

elasticity (the Cauchy-Maxwell equations). The sense in which the equations written in the variables $(m,\rho,C)$ and $(E,B)$ are Hamiltonian is especially interesting. For the corresponding structures for charged fluids and plasmas, see Marsden and Weinstein [1982] and Spencer [1982].

3. A deep understanding of the Hamiltonian formalism for incompressible fluids enabled Arnold [1966a,b] to prove the nonlinear stability of plane flows studied by Rayleigh in a situation where one would otherwise expect the usual difficulties with potential wells (Knops and Wilkes [1973], Marsden and Hughes [193, Sect. 6.6] and Ball and Marsden [1984]). It is hoped that a similar understanding in elasticity will shed light on the energy criterion.

4. Finally, we note that the Doyle-Ericksen formula is the spatial part of the stress-energy-momentum tensor that naturally arises when one couples elasticity to the gravitational field in Einstein's theory. The material picture is derived from this by choosing a reference body and slicing of spacetime relative to which the motion may be represented. Thus, in this sense, the Doyle-Ericksen formula may be regarded as the most basic of the four equivalent formulas

$$P = \rho_{Ref}\frac{\partial W}{\partial F}, \quad S = 2\rho_{Ref}\frac{\partial W}{\partial C}, \quad \sigma = 2\rho\frac{\partial e}{\partial g}, \text{ and } \Sigma = 2\rho\frac{\partial W}{\partial G}$$

## Appendix

### A Direct Verification of the Equivalence of the Stress Formulas

The components of the Green deformation tensor are defined by

$$\text{(1)} \qquad C_{AB} = F^a{}_A F^b{}_B g_{ab}$$

where $F^a{}_A$ and $g_{ab}$ represent the deformation gradient and spatial metric tensor, respectively. (The summation convention is assumed to hold except with respect to the arguments of functions.) By virtue of

(1), we may think of the $C_{AB}$'s as functions of the $F^a{}_A$'s and $g_{ab}$'s; viz.

$$(2) \qquad C_{AB} = \hat{C}_{AB}(F^a{}_A;\ g_{ab}) = \hat{C}_{AB}(F^1{}_1, F^1{}_2, \ldots;\ g_{11}, g_{12}, \ldots).$$

The physical interpretation of (2) goes as follows: The "strains" (i.e., $C_{AB}$'s) are functions of the gradients of the motion (i.e., $F^a{}_A$'s), and the length scales and angle measures of the ambient space, as manifested by the $g_{ab}$'s.

We will need to use the partial derivatives of $\hat{C}_{AB}$; namely

$$(3) \qquad \partial\hat{C}_{AB}/\partial F^b{}_C = \delta_{AC}F^a{}_B g_{ba} + \delta_{BC}F^a{}_A g_{ab}$$

$$(4) \qquad \partial\hat{C}_{AB}/\partial g_{ab} = F^a{}_A F^b{}_B$$

Let $W = W(C_{AB})$ be a given stored energy function and let us start by assuming that, say,

$$(5) \qquad S^{AB} = 2\rho_{Ref}\frac{\partial W}{\partial C_{AB}}$$

where $S^{AB}$ and $\rho_{Ref}$ represent the (symmetric) second Piola-Kirchhoff stress tensor and density in the reference configuration, respectively. A related potential, $\hat{W}$, may be defined by using (2);

$$(6) \qquad \hat{W}(F^a{}_A;\ g_{ab}) = W(\hat{C}_{AB}(F^a{}_A;\ g_{ab}))$$

Substituting (3) and (4) into (5) yields

$$(7) \qquad \rho_{Ref}\frac{\partial\hat{W}}{\partial F^a{}_A} = g_{ab}F^b{}_B S^{BA} = P_a{}^A$$

and

$$(8) \qquad 2\rho \frac{\partial \hat{W}}{\partial g_{ab}} = J^{-1} F^a{}_A S^{AB} F^b{}_B = \sigma^{ab}$$

where $\rho J = \rho_{Ref}$; $\rho$ is the density in the current configuration; J is the determinant of the deformation gradient; and $P_a{}^A$ and $\sigma^{ab}$ are the (unsymmetric) first Piola-Kirchhoff and Cauchy stress tensors, respectively.

## References


ABRAHAM, R.; MARSDEN, J.; and RATIU, T., 1983. Manifolds, Tensor Analysis and Applications. Reading, MA: Addison-Wesley Publishing Co., Inc.

ANTMAN, S. S., and OSBORNE, J. E., 1979. The principle of virtual work and integral laws of motion. Arch. Rat. Mech. An. 69:231-262.

ARNOLD, V., 1966. Sur la geometrie differentielle des groupes de Lie de dimension infinie et ses applications a l'hydrodynomique des fluids parfaits, Ann. Inst. Fourier, Grenoble 16:319-361.

ARNOLD, V., 1966b. An a priori estimate in the theory of hydrodynamic stability, Transl. Amer. Math. Soc. 79 (1969):267-269.

BALL, J. M., and MARSDEN, J. E., [1984]. Quasi-convexity, second variations and the energy criterion in nonlinear elasticity, Arch. Rat. Mech. An. (to appear).

BELINFANTE, F., 1939. On the current and the density of the electric charge, the energy, the linear momentum, and the angular momentum of arbitrary fields, Physica 6, 887 and 7, 449-474.

COLEMAN, B. D., and NOLL, W., 1959. On the thermostatics of continuous media, Arch. Rat. Mech. An. 4:97-128.

DOYLE, T. C., and ERICKSEN, J. L., 1956. Nonlinear Elasticity, in Advances in Appl. Mech. IV. New York: Academic Press, Inc.

HAWKING, S., and ELLIS, G., 1973. The Large Scale Structure of Spacetime, Cambridge, England: Cambridge Univ. Press.

HOLM, D., and KUPERSCHMIDT, B., 1983. Poisson brackets and Clebsch representations for magnetohydrodynamics, multifluid plasmas and elasticity Physica 6D, 347-363.

KNOPS, R., and WILKES, E., 1973. Theory of elastic stability, in Handbuch der Physik, VIa/3, C. Truesdell, ed., Berlin: Springer-Verlag.

MARSDEN, J. E., 1981. Lectures on Geometric Methods in Mathematical Physics, CBMS-NSF Regional Conference Series 37, Philadelphia, PA:SIAM.

MARSDEN, J. E., and HUGHES, T. J. R., 1983. Mathematical Foundations of Elasticity, Prentice-Hall.

MARSDEN, J. E.; RATIU, T.; and WEINSTEIN, A., 1984. Semidirect products and reduction in mechanics, Trans. Am. Math. Soc. 281, 147-177.

MARSDEN, J. E., and WEINSTEIN, A., 1982. The Hamiltonian Structure of the Maxwell-Vlasov equations, Physica 4D, 394-406.

MISNER, C.; THORNE, K.; and WHEELER, J., 1973. Gravitation. San Francisco, CA: W. H. Freman & Company, Publishers, Inc.

ROSENFELD, L., 1940. Sur le tenseur d'impulsion-énergie, Mem. Acad. R. Belg. Sci. 18, No. 6, 1-30

SIMO, J. C., and MARSDEN, J. E., [1983]. On the rotated stress tensor and the material version of the Doyle-Ericksen formula, Arch. Rat. Mech. An. (to appear).

SOKOLNIKOFF, I. S., 1956. The Mathematical Theory of Elasticity, (2d ed.) New York: McGraw-Hill Book Company.

SPENCER, R., 1982. The Hamiltonian structure of multi-species fluid electrodynamics, AIP Conf. Proc. 88, 121-126.

TRUESDELL, C., and NOLL, W., 1965. The Non-Linear Field Theories of Mechanics, Handbuch der Physic III/3, Berlin: ed. S. Flugge. Springer-Verlag.

# THE CAUCHY PROBLEM AND PROPAGATION OF SINGULARITIES

By Richard Melrose

This lecture was intended as an introduction to some of the recent progress in characterizing those linear partial differential operators which are hyperbolic in the sense that the Cauchy problem is locally well-posed in distributions or in $C^\infty$. Only single differential operators are considered, the same problems for determined systems are less well understood.

## 1. Hyperbolicity.

The Cauchy problem for a (partial) differential operator P, with respect to a hypersurface H, is the 'time' evolution problem. Namely, if m is the order of P, can a solution to Pu = 0 be uniquely specified by the first m terms in the Taylor series of u at H? A necessary condition for this, even in a formal sense, is that P be non-characteristic with respect to H. If $H = \{h=0\}$, where h is a $C^\infty$ defining function for H so $dh \neq 0$ on H, then P is non-characteristic with repsect to H at $\bar{x} \in H$ if

(1.1) $$P\, h^m(\bar{x}) \neq 0.$$

In any local coordinates $x = (x_1,\dots,x_n)$

(1.2) $$P = \sum_{|\alpha|\leq m} a_\alpha(x) D^\alpha$$

where we are assuming that the coefficients $a_\alpha$ are $C^\infty$ in some open region $\Omega \subset \mathbb{R}^n$ and in (1.2) the $D^\alpha$ are the differential monomials

$$D^\alpha = D_1^{\alpha_1} \dots D_n^{\alpha_n},\ D_j = -i\partial/\partial x_j,\ \alpha \in \mathbb{N},\ |\alpha| = \sum_{j=1}^{n} \alpha_j.$$

These coordinates can be chosen so that, at least near some particular $\bar{x} \in H$, $h = x_1$. The operator P can be written in the form

$$(1.3) \qquad P = \sum_{k=1}^{m} P_{m-k}(x,D')\, D_1^k$$

where $P_j$ is a differential operator in $x' = (x_2,\ldots,x_n)$ of order j, though with coefficients depending on x. The non-characteristic condition (1.1) is then just $P_0(\bar{x}) \neq 0$, since this is a differential operator of order 0, i.e. a function.

The classical form of the Cauchy problem is the search for a solution $u = u(x)$ to

$$(1.4) \qquad \begin{cases} Pu = 0 \text{ in } \Omega_0 = \{x \in \Omega;\ x_1 > 0\} \\ D_1^j u(0,x') = u_j(x') \text{ on } H \end{cases}$$

where the $u_j$ are prescribed. Naturally, the region $\Omega$ in which the equation is to hold must be specified and either an acceptable degree of regularity for u prescribed or else some other consistent meaning must be given to the boundary condition in (1.4). Notice that (1.1) is actually a necessary and sufficient condition for the solvability of (1.4) in the formal category of power series in $x_1$ with coefficients which are $C^\infty$ functions at $\bar{x}' \in H$, $\bar{x} = (0,\bar{x}')$. The same is true in the analytic category, assuming in particular that the coefficients are analytic (the Cauchy-Kowalevskia Theorem). On the other hand, (1.1) is far from sufficient to guarantee the solvability of (1.4) near $\bar{x}$ for general $C^\infty$ data.

One elegant way of dealing with the boundary conditions is to abolish them completely and pass to Schwartz's distributional formulation of the Cauchy problem. If $f \in C^{-\infty}(\Omega)$ is a distribution in $\Omega$, or just a function in $\Omega$, which vanishes in $x_1 < 0$, one seeks a distribution $u \in C^{-\infty}(\Omega)$ satisfying

$$(1.5) \qquad Pu = 0 \text{ in } \Omega,\ u = f = 0 \text{ in } x_1 < 0.$$

Thus, the forcing term f is zero in the 'past' and one looks for the solution 'caused' by that forcing, i.e. also zero in the past. In fact (1.4) and (1.5) are very closely related. In each case well-posedness is the requirement of unique solvability of the problem for a stated general class of data. If each formulation is strengthened by requiring well-posedness for each of a family of initial hypersurfaces $x_1 = t$, $|t| < \delta, \delta > 0$, then they are equivalent. The passage from (1.4) to (1.5) is duHamel's principle and the reverse implication is accomplished by Schwartz's weak formulation of the boundary condition and Peetre's partial hypoellipticity theorem.

We shall therefore adopt as the definition of hyperbolicity a suitable form of the well-posedness of (1.5).

(1.6) **DEFINITION** (Ivrii). If P is a linear differential operator of order m with $C^\infty$ coefficients on the manifold X and $t \in C^\infty(X)$ is a real-valued function then P is said to be hyperbolic at $\bar{x} \in X$ with respect to t if

(1.7) $$P\, t^m(\bar{x}) \neq 0$$

(1.8) $$\begin{cases} \bar{x} \text{ has a neighborhood } \Omega \subset X \text{ such that} \\ \text{for some } \epsilon>0, \text{ if } D_\tau = \{v \in C^{-\infty}(\Omega);\\ v=0 \text{ in } t<t(\bar{x})+\tau\} \text{ and } |\tau|<\epsilon, \text{ then} \\ P: D_\tau \to D_\tau \text{ is an isomorphism.} \end{cases}$$

Here, (1.7) is just the non-characteristic requirement (1.1). It is important to note that the assumption of locality in (1.8), that $\Omega$ is some possibly very small neighborhood of $\bar{x}$, is weak as far as the existence is concerned but is a strong assumption as it applies to uniqueness. Thus (1.8) amounts to the imposition of a very weak form of the finite speed of propagation condition which is fundamental to the notion of hyperbolicity.

The basic problem of interest here is: Which operators are hyperbolic? One would like to find a characterization of hyperbolicity directly in terms of the coefficients in (1.2). In this generality the problem is somewhat daunting, although some necessary conditions are described below. We shall focus on the simpler question of characterizing strong hyperbolicity.

**(1.9) DEFINITION.** If P is a differential operator of order m on a manifold X and $t \in C^{\infty}(X)$ is real-valued then P is said to be strongly hyperbolic at $\bar{x}$ with respect to t if P + Q is hyperbolic at $\bar{x}$ with respect to t for every $C^{\infty}$ differential operator Q of order m−1 (or less).

In this definition the region $\Omega = \Omega(Q)$ in which the Cauchy problem is well-posed for P + Q may depend, a priori, on Q. It is a consequence of the characterization of strong hyperbolicity discussed below that the region can be chosen independent of Q. Moreover, if P is strongly hyperbolic at $\bar{x}$ with respect to t then $dt \in T^*_{\bar{x}}X$ has a neighborhood $\Upsilon$ such that P is strongly hyperbolic with respect to $t' \in C^{\infty}(X)$, at $\bar{x}$, if $dt'(\bar{x}) \in \Upsilon$.

## 2. Constant coefficients.

The one case in which hyperbolicity as in Definition 1.6, has been completely characterized is that in which the coefficients of P in (1.2) are constants; it is then natural, although by no means necessary, to take t to be a linear function. Now,

$$P = p(D) = \sum_{|\alpha| \leq m} a_\alpha D^\alpha, \tag{2.1}$$

is a polynomial in the indeterminates $D_1,\ldots,D_n$ and hyperbolicity is a condition on the characteristic variety of P, which is just the algebraic variety

$$V_P = \{\zeta \in \mathbb{C}^n;\ p(\zeta) = 0\}.$$

Namely,

$$\left\{\begin{array}{l} \text{There exists } T>0 \text{ such that if } \xi\in\mathbb{R}^n, \\ \tau\in\mathbb{C} \text{ satisfy } p(\xi+\tau dt)=0 \text{ then } |\mathrm{Im}\ \tau|<T. \end{array}\right. \tag{2.2}$$

**(2.3) THEOREM (Gårding).** If P has constant coefficients then it is hyperbolic at $\bar{x}$ with respect to t if and only if (1.7) and (2.2) hold.

If t is linear then the condition of hyperbolicity is independent of the point $\bar{x}$. The necessity of the algebraic condition (2.2) is proved by examining the exponential solutions of $Pu = 0$, $u = \exp(ix.\zeta)$ where $\zeta \in V_P$, the sufficiency by use of harmonic analysis, i.e. the Fourier-Laplace transform, to construct a fundamental solution $E \in C^{-\infty}(\mathbb{R}^n)$,

$$PE = \delta(x),\ E = 0 \text{ in } x_1 < 0. \tag{2.4}$$

E actually has its support in a proper closed conic subset K of $x_1 \geq 0$ and this allows (1.4) and (1.5) to be solved by convolution. The regions $\Omega$ in which the Cauchy problem is well-posed are limited by a condition of P-convexity,

$$(-K + \{x\}) \cap \{x_1 \geq 0\} \subset \Omega\ \forall\ x \in \Omega,\ t = x_1. \tag{2.5}$$

When P is given by (2.1) the principal part of P is given by

(2.6) $$P_m = p_m(D) = \sum_{|\alpha|=m} a_\alpha D^\alpha.$$

From (2.2) it follows that if P is hyperbolic with respect to t then so is $P_m$. For such a homogeneous polynomial satisfying (1.7) the condition (2.2) is equivalent to

(2.7) If $\xi \in \mathbb{R}^n$ then $p_m(\xi + \tau dt) = 0$ has only real roots $\tau$.

**(2.8) DEFINITION.** P(D), given by (2.1), is said to be strictly hyperbolic with respect to t if (1.7) holds and the roots of $p_m(\xi + \tau dt) = 0$ are real and distinct whenever $\xi \in \mathbb{R}^n \setminus \mathbb{R}dt$.

**(2.9) THEOREM.** If P has constant coefficients then P is strongly hyperbolic with respect to t if and only if it is strictly hyperbolic with respect to t.

The point of keeping the two notions, of strict hyperbolicity for the algebraic condition and of strong hyperbolicity for the analytic condition, separate is that they are indeed distinct for variable coefficients operators. The algebraic condition (2.2) can also be replaced by inequalities on the lower order terms in P, once it is known that the principal part is hyperbolic.

**(2.10) THEOREM (Svensson).** If P(D) satisfies (1.7) and has principal part $P_m(D)$ hyperbolic with respect to t then P is hyperbolic with respect to t if and only if there exists C > 0 such that

$$|p(\xi)| \leqslant C \sum_\alpha |D^\alpha p_m(\xi)| \quad \forall\ \xi \in \mathbb{R}^n.$$

## 3. Strict hyperbolicity.

When P is an operator with constant coefficients it is natural to restrict coordinate transformations to be linear so as to preserve this condition. When P has variable coefficients this is no longer the case so we shall view P as acting on functions, or distributions, on a $C^\infty$ manifold X, even though all considerations will be local near some point $\bar{x}$. Then (2.2) is the local coordinate form of P. The principal symbol of P, given by (2.2) is

$$p_m(x,\xi) = \sum_{|\alpha|=m} a_\alpha(x)\xi^\alpha \tag{3.1}$$

precisely as in (2.6). Since the indeterminant D is transformed under the Jacobian by a coordinate transformation $p_m$ becomes an invariantly defined function on the cotangent bundle $T^*X$ when, in any coordinate system, $(x,\xi)$ is identified with the 1-form $\xi dx$.

(3.2) **DEFINITION.** P is said to be strictly hyperbolic at $\bar{x} \in X$ with respect to $t \in C^\infty(X)$ if for each $x \in \Omega$, some neighborhood of $\bar{x}$, $p_m(x,\cdot)$ is a strictly hyperbolic polynomial with respect to $dt(x)$.

(3.3) **THEOREM (Petrowskii, Leray).** If P is strictly hyperbolic with respect to t at $\bar{x}$ then P is strongly hyperbolic with respect to t at $\bar{x}$.

This result, and the elucidation of its proof, has been a central point in the development of Fourier integral operators. Let us briefly describe a proof by factorization, using pseudodifferential operators. First introduce local coordinates in which $t = x_1$ and $\bar{x}$ is the origin. By examining the properties of strictly hyperbolic polynomials it is easy to extend P, from its definition in some small neighborhood of 0, to an operator on the whole of $\mathbb{R}^n$, again denoted P, which is everywhere strictly hyperbolic with respect to $x_1$ and has constant coefficients in $|x| > \epsilon$. The calculus of pseudodifferential operators allows one to factorize P:

(3.4) $$P = a(x)(D_1 - A_1(x,D'))\dots(D_1 - A_m(x,D')) + \sum_{j=1}^{m-1} R_j(x,D')d_1^j.$$

Here the $A_j$ are pseudodifferential operators in $x' = (x_2,\dots,x_n)$ with compact supports and depending smoothly on $x_1$ and the error terms are similar smoothing operators in x'. Such a factorization (3.4) easily leads to energy estimates. If $\phi \in C_c^\infty(\mathbb{R}^n)$ has support in $x_1 \geqslant 0$ then for some constant C and all $T \leqslant \bar{T}(P) > 0$,

(3.5) $$\sum_{|\alpha|\leqslant m-1} \|D^\alpha \phi\| \leqslant CT\ \|P\phi\|$$

where the norms are $L^2$ norms over the set $x_1 < T$. Using functional anaytic arguments, and commutator arguments, the well-posedness of the Cauchy problem for the original P follows.

For the solvability of (1.5) when P is strictly hyperbolic $\Omega$ must still satisfy a P-convexity condition. This condition is given in terms of the bicharacteristics of P. It can always be assumed that $p_m$ is real-valued and then the integral curves on the surface $p_m = 0$ of the vector field in $T^*X$

(3.6) $$H_{p_m} = \sum_{j=1}^{n} \partial_{\xi_j} p_m(x,\xi)\partial_{x_j} - \partial_{x_j} p_m(x,\xi)\ \partial_{\xi_j},$$

the Hamilton vector field of P, are called the bicharacteristics of P. The condition on $\Omega$ for well-posedness of the Cauchy problem (1.5) is:

(3.7) For each $x\in\Omega$ and $\xi\in T_x^*X\setminus 0$ with $p_m(x,\xi)=0$, $(x,\xi)$ is the endpoint of precisely one segment $b(x,\xi)$ of bicharacteristic in $T_*\Omega$ with $x_1 = 0$ at the other endpoint.

For each $x \in \Omega$ the union of the projections into $\Omega$ of the bicharacteristic segments $b(x,\xi)$ forms a closed set $\Gamma(x) \subset \Omega$, the domain of dependence of x. The finite speed of propagation can then be expressed concretely as:

(3.8) If u,f satisfy (1.5) with P strictly hyperbolic with respect to $x_1$ and $\Omega$ P-convex as in (3.7) then for each $x \in \Omega$, $f=0$ in a neighborhood of $\Gamma(x)$ implies $u=0$ in a neighborhood of $\Gamma(x)$.

The same conclusion holds for the singularities of the solution of (1.5); if f is $C^\infty$ in a neighborhood of $\Gamma(x)$ then so is u. A more refined version of this is the celebrated theorem on the propagation of singularities. For this one needs to microlocalize the notion that u is singular at x, i.e. that u is not $C^\infty$ in a neighborhood of x, to that of the wavefront set of u. We do not give a precise definition here but simply note that if $u \in C^{-\infty}(X)$, is a distribution, then WF(u) $\subset T^*X\setminus 0$ is closed and if $(x',\xi') \notin WF(u)$ then near x' u can be considered as a $C^\infty$ function in any direction sufficiently near to $\xi'$, with distributional values in the other variables.

**(3.9) THEOREM (Hörmander).** Suppose that P is strictly hyperbolic with respect to $x_1$ and $\Omega$ is a P-convex set in the sense of (3.7). Then if u, $f \in C^{-\infty}(\Omega)$ satisfy (1.5), $WF(u)\setminus WF(f) \subset \{p_m(x,\xi) = 0\}$ and if $b(x,\xi) \cap WF(f) = \emptyset$ then $(x,\xi) \notin WF(u)$, when $p_m(x,\xi) = 0$.

There are less precise notions of wavefront set than WF(u), such as $WF_s(u)$ which contains only the points $(x,\xi) \in T^*X\setminus 0$ near which u is not, in a suitable microlocal sense, in the Sobolev space $H^s_{loc}(X)$. There is a version of Theorem 3.9 relative to his type of wavefront set. Thus, if P and $\Omega$ are as before and u,f satisfy (1.5) then

(3.10) $p_m(x,\xi) \neq 0$ implies that $(x,\xi) \in WF_{m+s}(u)$ if $(x,\xi) \in WF_s(f)$

and much more significantly,

(3.11) if $p_m(x,\xi) = 0$ then $b(x,\xi) \cap WF_{s+1}(f) = \emptyset$ implies $(x,\xi) \notin WF_{m+s}(u)$.

From such a result on the propagation of singularities a weak form of the energy estimates (3.5) can be recovered, almost but not quite enough to show the hyperbolicity of P directly. We shall return to this point below.

Finally we note that the inequalities (3.5), for all compactly supported functions vanishing in $x_1 < -\delta$, $\delta > 0$, are actually equivalent to the strict hyperbolicity of P. Strongly hyperbolic operators which are not strictly hyperbolic satisfy similar estimates in which only lower order (possibly only negative order) derivatives of $\phi$ can be bounded in $L^2$ by the $L^2$ norm of $P\phi$. Correspondingly solutions of such equations are in general less regular than the solutions of strictly hyperbolic equations.

4. **Effective hyperbolicity.**

The first and basic result on hyperbolic operators with variable coefficients is:

**(4.1) THEOREM (Lax, Mizohata).** If P is hyperbolic at $\bar{x}$ with respect to t then there is a neighborhood $\Omega$ of $\bar{x}$ such that for each $x \in \Omega$, $p(x,\cdot)$ is a hyperbolic polynomial with respect to $dt(x)$.

Following (2.8), P is strictly hyperbolic at $\bar{x}$ if $p_m(\bar{x},\cdot)$ is strictly hyperbolic with respect to $dt(\bar{x})$. Thus, we shall consider the case of multiple characteristics; $\rho = (x,\xi) \in M = T^*X\setminus 0$ is a characteristic of order k if

$$D^\beta p_m(\rho) = 0,\ |\beta| < k,\ D^\beta p_m(\rho) \neq 0 \quad (4.2)$$
$$\text{for some } |\beta| = k,$$

where $D = (D_{x_1},\dots,D_{x_n},D_{\xi_1},\dots,D_{\xi_n})$. In fact if P is hyperbolic at $\bar{x}$ with respect to $x_1$ then at any characteristic of order k, $\rho \in T^*_{\bar{x}}X\setminus 0$, $\partial^k_{\sigma_1} p_m(\rho) \neq 0$. As a consequence of this and the Malgrange preparation theorem $p_m$ takes the form

$$p_m(x,\xi) = q(x,\xi) \sum_{j=0}^{k} q_{k-j}(x,\xi')\xi_1^j,\ q_0 = 1, \quad (4.3)$$

near a characteristic $\rho$ of order k, with $q(\rho) \neq 0$ and the $q_p$ homogeneous of degree p in $\xi' = (\xi_2,\dots,\xi_n)$. Using Boutet de Monvel's pseudodifferential operator analogue of the preparation theorem P itself becomes, microlocally

$$P \equiv Q(x,D_x)\Big(\sum_{j=0}^{k} Q_j(x,D')D_1^j\Big) \quad (4.4)$$

near $\rho = (\bar{x},\bar{\xi})$, i.e. a differential operator in $D_1$, non-characteristic with respect to $x_1$ but with pseudodifferential operator coefficients.

The important ideas of Ivrii and Petkov, leading to necessary conditions for hyperbolicity, show that if P is strongly hyperbolic at $\bar{x}$

with respect to t then for some neighborhood $\Omega$ of $\bar{x}$, $p_m$ has at most double characteristics in $T^*\Omega\setminus 0$.

Multiple characteristics correspond to zeros of the Hamilton vector field $H_{p_m}$, the further classification of operators with double points reduces to the discussion of the form of the zero of $H_{p_m}$. On $T^*X$ there is a natural 2-form, given in any canonical coordinates $(x,\xi)$ by

$$\omega = \sum_{j=1}^{n} d\xi_j \wedge dx_j.$$

It is this symplectic structure which fixes the Hamilton vector field given by (3.6),

$$(4.5) \qquad \omega(H_{p_m},\cdot) = -\, dp_m(\cdot).$$

In view of this the linearization of $H_{p_m}$ at a multiple characteristic $\rho$ is determined by the Hessian of $p$, at $\rho$. The linearization is the map

$$(4.6) \qquad T_\rho M \ni V \mapsto F_\rho V \in T_\rho M, \; M = T^*X\setminus 0,$$

where $F_\rho V$: $T^*_\rho M \ni df \mapsto VH_{p_m}f(\rho)$. This is also called the fundamental matrix.

**(4.7) LEMMA.** If P is hyperbolic at $\bar{x}$ with respect to t and if $\rho$ is a sufficiently small neighborhood of $\bar{x}$ then at any double characteristic $\rho \in T^*X\setminus 0$ of P, $F_\rho$ has eigenvalues $\pm e, \pm i\lambda_1, \ldots, \pm i\lambda_r; e, \lambda_1, \ldots, \lambda_r \in \bar{\mathbb{R}}^+$.

**(4.8) DEFINITION.** If P is hyperbolic at $\bar{x}$ with respect to t then P is effectively hyperbolic at a multiple characteristic $\rho \in T^*_{\bar{x}}X\setminus 0$ if $F_\rho$ has a non-zero real eigenvalue.

Now we are in a position to give the main characterization result.

(4.9) **THEOREM (Ivrii & Petkov, Hörmander, Iwasaki, Melrose, Nishitani).** P is strongly hyperbolic at $\bar{x}$ with respect to t if and only if its principal part $p_m(x,\cdot)$ is hyperbolic with respect to dt(x), for all x in some neighborhood of $\bar{x}$, and in $T^*_{\bar{x}}X\backslash 0$ P is effectively hyperbolic at each of its multiple characteristics.

The necessity of effective hyperbolicity for strong hyperbolicity is discussed briefly below, we shall now comment on the proof of sufficiency. If P is effectively hyperbolic at the double characteristic $\rho \in T^*X\backslash 0$ then under a homogeneous, linear, symplectic transformation of $T^*_\rho M$, $p_m$ can be reduced to the form

$$p_m(x,\xi) = q(x,\xi)(\xi_1^2 - x_1^2\xi_n^2 - b(x',\xi')),\ b \geqslant 0, \tag{4.10}$$

where $\rho = (0,\tilde{\xi})$, $\tilde{\xi} = (0,\dots,0,1)$. By a related Fourier integral operator P can be reduced to a special form

$$P = D_1^2 - x_1^2H(x,D') - B(x',D') \text{ microlocally near } \rho \tag{4.11}$$

where B and H are pseudodifferential operators with principal symbols, $b > 0$, $h > 0$. Using this and energy estimates, the propagation of singularities for P can be analysed. If $p_m(x,\cdot)$ is hyperbolic for $x \in \Omega$, a neighborhood of $\bar{x}$, and $p_m(\bar{x},\cdot)$ has only effectively hyperbolic double characteristics, consider all the segments of (unparametrized) bicharacteristic in $T^*\Omega$ along which $x_1$ is decreasing. For any $(x,\xi) \in T^*\Omega\backslash 0$, with $p_m(x,\xi) = 0$ consider any curve b which is a union of closures of such bicharacteristic segments, which has initial point $(x,\xi)$ and terminal point in $x_1 = 0$ and along which $x_1$ is strictly decreasing. Then set

$$\Gamma(x,\xi) = \cup\, b,\ p_m(x,\xi) = 0 \tag{4.12}$$

the union of all such curves. For a suitably small $\Omega$, P-convex $\Gamma(x,\xi)$ is closed in $T^*\Omega\setminus 0$ and if u,f satisfy (1.5) then

$$\Gamma(x,\xi) \cap WF(f) = \emptyset \text{ implies } (x,\xi) \notin WF(u). \tag{4.13}$$

Thus, $\Gamma(x,\xi)$ is the microlocal (past) dependence domain of $(x,\xi)$. If $\Omega$ is chosen small enough, and P-convex, then there is at most one double characteristic $\rho \in \Gamma(x,\xi)$. Moreover, as in the relative version (3.11) of Theorem 3.9 there is a constant, N, measuring the loss of Sobolev regularity:

$$\Gamma(x,\xi) \cap WF_{s+1+N}(f) = \emptyset \text{ implies } (x,\xi) \notin WF_{m+s}(u), \tag{4.14}$$

when u,f satisfy (1.5). The same methods lead to a weakened form of the energy estimates (3.5),

$$\|\phi\|_{m+s} \leq C(\|P\phi\|_{s+1+N} + \|\phi\|_{-M}) \tag{4.15}$$

for all $\phi \in C_c^\infty(\Omega)$ with support in $x_1 > -\delta$, with the norms Sobolev norms over the region $x_1 < T$. Here, the error term $\|\cdot\|_{-M}$ is an arbitrarily low order Sobolev norm. As noted in section 3, these estimates do not lead directly to the required existence theorem. However, using a suitable approximation argument and the existence result for strictly hyperbolic operators, (4.15) does lead to existence and uniqueness results modulo finite kernels and cokernels. To prove estimates (4.15) without the error term on the right it is necessary to carry through the reduction and estimates for an operator depending on a parameter. The microlocal dependence domains also give the full dependence domains: if

$$\Gamma(x) = \bigcup_{\Xi\in T^*_xX\setminus 0} \Gamma(x,\xi) \tag{4.6}$$

then (3.8) holds when P is stongly hyperbolic.

The necessity of effective hyperbolicity in Theorem 4.9 is a special case of a more general condition for hyperbolicity. At any double characteristic of P, $\rho \in T^*X\setminus 0$, the subprincipal symbol of P is well defined by reference to any local coordinates x, as in (1.2):

$$\sigma ub(P)(\rho) = \sum_{|\alpha|=m-1} a_\alpha(x)\xi^\alpha + \frac{i}{2}\sum_{j=1}^{n} \partial_{x_j}\partial_{\xi_j}p_m(x,\xi). \tag{4.17}$$

If P is hyperbolic then the following Lemma 4.7 we set

$$\mathrm{tr}^+(P)(\rho) = \sum_p \lambda_p \text{ at any multiple characteristic.} \tag{4.18}$$

**(4.19) THEOREM (Ivrii & Petkov, Hörmander).** If P is hyperbolic at $\bar{x}$ with respect to t and $\rho \in T^*_{\bar{x}}X\setminus 0$ is a multiple characteristic which is not effectively hyperbolic then $\sigma_{sub}(P)(\rho)$ is real and $|\sigma_{sub}(P)(\rho)| \leqslant \frac{1}{2}\,\mathrm{tr}^+(P)(\rho)$.

These condition are not sufficient to guarantee hyperbolicity even for operators with only double characteristics and we propose a strengthening as follows. Assuming that $p_m(x,\cdot)$ is hyperbolic with respect to dt(x) for $x \in \Omega$, let $\Sigma_E \subset T^*\Omega\setminus 0$ be the set of effectively hyperbolic characteristis. For $\rho \in \Sigma_E$ let $e(\rho)$ be the positive eigenvalue of $R_\rho$ and set

$$s(\rho) = |\mathrm{IM}\ \sigma_{sub}(P)(\rho)| + \inf\{|\mathrm{Re}\ \sigma_{sub}(P)(\rho) - s|;\ |s| \leqslant \tfrac{1}{2}\mathrm{tr}^+(P)(\rho)\},$$

measuring the distance of the subprincipal symbol from the real segment $|s| \leqslant \mathrm{tr}^+(P)(\rho)$.

**(4.21) CONJECTURE.** If P is hyperbolic at $\bar{x}$ with respect to t, then $s(\rho)/e(\rho)$ is uniformly bounded on $\Sigma_E \cap T^*\Omega\setminus 0$ for some neighborhood $\Omega$ of $\bar{x}$.

## 5. References.

There is a very large number of research papers examining various aspects of the hyperbolicity of linear differential operators. We shall only give here references to cover the results discussed above.


1. V.Ya. Ivrii, Sufficient conditons for regular and completely regular hyperbolicity. Trudy Moskov. Obsč. 33 (1975), 1-65.

2. L. Hörmander, Linear Partial Differential Operators. Springer-Verlag, New York, 1969.

L. Svensson, Necessary and sufficient conditions for the hyperbolicity of polynomials with hyperbolic principal part. Ark. Mat. 8 (1969), 145-162.

3. F. Treves, Introduction to Pseudodifferential and Fourier Integral Operators. Plenum, New York, 1980.

M. Taylor, Pseudodifferential Operators. Princeton University Press, Princeton, 1980.

J.J. Duistermaat & L. Hörmander, Fourier integrals operators II. Acta Math. 128 (1972), 183-269.

4. V.Ya. Ivrii & V.M. Petkov, Necessary conditions for the correctness of the Cauchy problem for non-strictly hyperbolic equations, Uspechi Mat. Nauk 29 (1974), 3-70.

L. Hörmander, The Cauchy problem for differential equations with double characteristics. J. d'analyse math. 32 (1977), 188-196.

N. Iwasaki, Cauchy problems for effectively hyperbolic equations (preprint).

R.B. Melrose, The Cauchy problem for effectively hyperbolic operators. To appear.

# ANALYTICAL THEORIES OF VORTEX MOTION

John Neu

## 1. Fluid mechanics as the dynamics of vortices.

In an incompressible Newtonian fluid, the physical principles of mass and momentum conservation lead directly to the Navier Stokes equations for the velocity $\underline{u}$ and pressure p:

$$\underline{\nabla}\cdot\underline{u} = 0, \quad \frac{D\underline{u}}{Dt} = -\underline{\nabla}p + \nu\Delta\underline{u}, \quad \frac{D}{Dt} \equiv \partial_t + \underline{u}\cdot\underline{\nabla}. \tag{1.1}$$

We pursue an alternative formulation which views fluid mechanics as the dynamics of the vorticity

$$\underline{w} \equiv \underline{\nabla} \times \underline{u}. \tag{1.2}$$

The motivation for this view is that instablities leading to turbulence manifest themselves through the rearrangement of smoothly distributed vorticity in the initial laminar flow into concentrated vorticies characteristic of turbulent flow. The ongoing dynamical process of turbulence is the interaction of many vortices.

From the Navier-Stokes equations, we may derive a dynamical equation for the vorticity. Taking the curl of the momentum balance equation gives

$$\frac{D\underline{w}}{Dt} = (\underline{w}\cdot\underline{\nabla})\underline{u} + \nu\Delta\underline{w}. \tag{1.3}$$

In order that (1.3) be a closed equation for $\underline{w}$, we must give $\underline{u}$ as a functional of $\underline{w}$. The general form of a velocity field $\underline{u}$ with a prescribed vorticity distribution $\underline{w}$ in unbounded fluid which satisfies $\int |\underline{w}|^2\, dV < \infty$ is

$$\underline{u}(\underline{x}) = -\frac{1}{4\pi}\int \frac{(\underline{x} - \underline{x}') \times w(\underline{x}')}{|\underline{x} - \underline{x}'|^3}\, d\underline{x}^3 + \underline{u}_e(\underline{x}). \tag{1.4}$$

The integral term in (1.4) represents the velocity induced solely by the vorticity distribution, and $\underline{u}_e$ is the irrotational velocity produced by distant influences. In practice, the distant influences may be other vorticies, or the solid boundaries of an experimental apparatus.

## 2. Two dimensional vortex motion.

If the flow is two dimensional with

$$\underline{u} = u(x,y)\hat{x} + v(x,y)\hat{y}, \tag{2.1}$$

then the vorticity is essentially a scalar:

$$\underline{w} = w(x,y)\hat{z}, \qquad w \equiv v_x - u_y \tag{2.2}$$

In this case, the vortex dynamics equations (1.3), (1.4) for $\underline{u}_e = 0$ reduce to

$$w_t + uw_x + vw_y = \nu \Delta w \tag{2.3}$$

$$u(z) - iv(z) = -\frac{i}{2\pi}\int \frac{w(z')}{z-z'}\,dx'\,dy', \tag{2.4}$$

where $z \equiv x + iy$, $z' = x' + iy'$.

The physical content of (2.3), (2.4) is that the scalar vorticity is convected by the self induced velocity and diffused by viscosity.

## 3. A necessary criterion for instability of parallel flows.

A fundamental problem in two dimensional vortex motion is the development of vorticial structures in the layer of fluid between two streams which travel at different velocities.

The initial growth of the vortex structure is due to a hydrodynamic instability. Lord Rayleigh considered perturbations of the parallel flow

$$\underline{u} = U(y)\hat{x} \tag{3.1}$$

in an inviscid fluid. He showed that a necessary condition for temporal instability of a spatially periodic vorticity perturbation is that the velocity profile $U(y)$ have an inflexion point $y = y_0$ where $U''(y_0)$ changes sign. The vorticity distribution of the parallel flow is

$$w = - U_y. \tag{3.2}$$

If $U''(y)$ changes sign at $y = y_0$, then the vorticity has a local extremum at the same level. This suggests that layers of concentrated vorticity will be hydrodynamically unstable.

## 4. Vortex sheets and the Birchoff equation.

To capture the essence of this instability, we consider the most singular case, in which all the vorticity is concentrated in an infinitely thin vortex layer. We think of the vortex layer as a continuous distribution of infinitesimal point vorticies along a curve. The curve is given in parametric form

$$z = z(\gamma,t) = x(\gamma,t) + iy(\gamma,t), \tag{4.1}$$

where $\gamma$ is the total circulation around the portion of vortex layer between $z = z(\gamma,t)$ and a given reference vortex at $z = z(0,t)$.

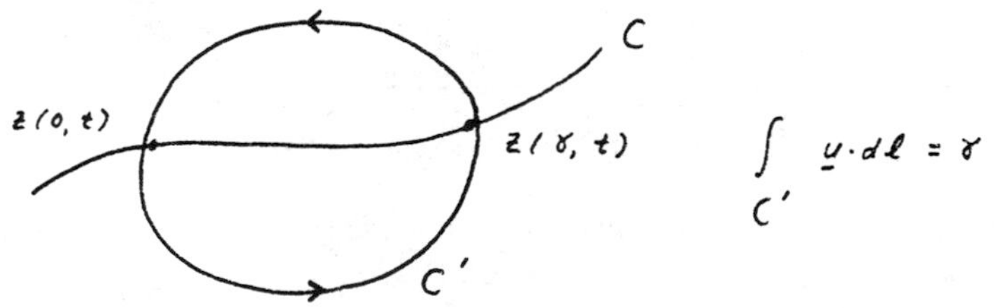


Figure 1

For fixed values of $\gamma$, (4.1) represents the path of individual vorticies. The velocity of any given vortex is the superposition of all the velocities generated by its neighbors. In mathematical terms, this statement reads

$$\bar{z}_t(\gamma,t) = -\frac{i}{2\pi}\int_C \frac{d\gamma'}{z(\gamma,t) - z(\gamma',t)}. \tag{4.2}$$

(4.2) is known as the Birchoff equation for vortex sheet dynamics [1].

A uniform vortex sheet with circulation $\sigma$ per unit length is given by

$$z(\gamma) = \frac{\gamma}{\sigma}. \tag{4.3}$$

To consider perturbations of the uniform vortex sheet, we take

$$z(\gamma,t) = \frac{\gamma}{\sigma} + \xi(\gamma,t). \tag{4.4}$$

The equation for $\epsilon(\gamma,t)$ is

$$\bar{\xi}_t = \frac{i\sigma^2}{2\pi}\int\left[\frac{\xi - \xi'}{(\gamma - \gamma')^2} - \sigma\frac{(\xi - \xi')^2}{(\gamma - \gamma')^3} + \cdots\right]d\gamma'. \tag{4.5}$$

The linearized equation is obtained by retaining only the first term in the integrand of (4.5). The linearized equation has solutions

$$\xi(\gamma,t) = e^{\frac{3}{4}\pi i}\, e^{\frac{\sigma k t}{2}} \cos\frac{k\gamma}{\sigma} \tag{4.6}$$

which represent temporally growing modes with growth rate $\alpha = \sigma k/2$ and spatial period $\lambda = 2\pi/k$. Figure 2 depicts the vortex trajectories according to linearized solution (4.6). The vortices migrate from their initial positions close to the x axis along 45° lines. We see how this migration leads to alternating concentration and rarefaction of vorticity along the length of the sheet.

The vortex dynamics predicted by the Birchoff equation (4.2)

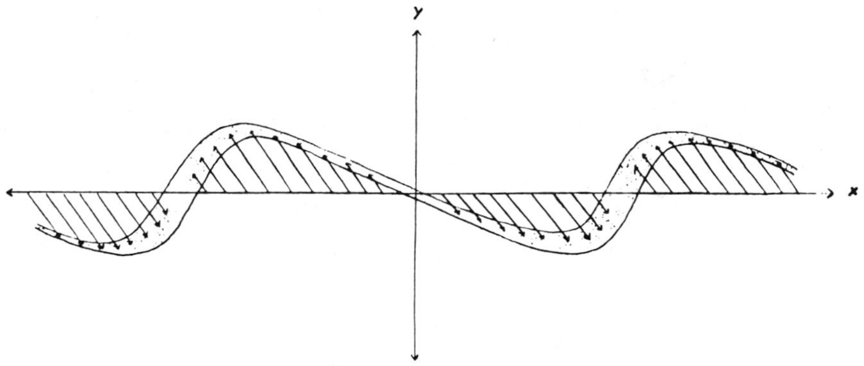


Figure 2

cannot be a dynamically consistent idealization of real flows for all times. From (4.6), we see that the growth rate $\alpha$ of a periodic ripple in the vortex sheet with wavelength $2\pi/k$ is

$$\alpha = \frac{\sigma k}{2} \tag{4.7}$$

according to linear theory. Hence the linearized evolution from an initial condition $\xi = \xi'(\gamma,0)$ which is analytic in a finite strip about the real axis of the $\gamma$ plane will develop a singularity in finite time.

Investigations [2], [3] of the full Birchoff equation suggest that singularities form from even simple sinusoidal perturbations of the uniform vortex layer. The nonlinearity of the Birchoff equation produces the higher harmonics necessary to generate the singularity. In [2], D. Moore performs an asymptotic analysis of the vortex sheet evolution resulting from the initial condition

$$z(\gamma,0) = \gamma + i\epsilon \sin \gamma \tag{4.8}$$

in the limit $\epsilon \longrightarrow 0$. He presents evidence that a singularity located at $\gamma = 0$ appears at time

$$t = t_c \simeq 2 \ln \frac{4}{\epsilon} \tag{4.9}$$

and that the $n^{th}$ harmonic coefficient $a_n$ of the singular configuration at $t = t_c$ has magnitude

$$a_n = 0(n^{-5/2}). \tag{4.10}$$

The vortex layer is only slightly deformed from the horizontal, while the vortex strength density $\sigma$ as a function of arclength s from the singularity has a cusp of the form

$$\sigma(s) \sim \sigma_0 - \sigma_1 s^{1/2} + \cdots. \tag{4.11}$$

The above results, though strongly suggested, are not

conclusive because the estimate of the $n^{th}$ harmonic coefficient $a_n(t)$ at time t is valid only for $1 << n << t$, and not uniformly as $n \longrightarrow \infty$. Indeed, a new attack on the problem by D. Merion, G. Baker, S. Orzag in [3], based on extrapolation of Taylor series in time, found a slightly weaker singularity that forms slightly later. In their analysis, they found that singular configuration at breakdown has $a_n = 0(n^{-2.8})$.

## 5. Large vorticial structures in shear layers.

In real flows, there are finitely thick vortex layers in viscous fluid. The wavenumbers of unstable modes are limited to a finite band about $k = 0$. We consider the generation of vorticial structures from perturbations of a monotonically increasing velocity profile $U(y)$ which asymptotes to $\pm U_\infty$ as $y \longrightarrow \pm \infty$.

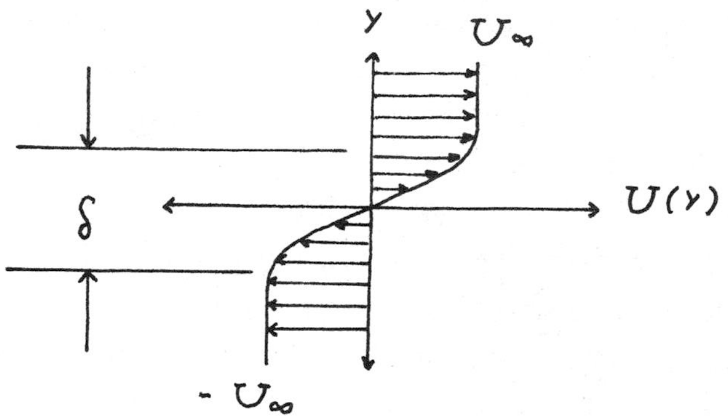


Figure 3

The length

$$\delta \equiv \frac{2U_\infty}{\max(U'(y))} \tag{5.1}$$

provides the measure of the vortex layer thickness. $\delta$ is the vertical extent of a layer with uniform vorticity $u = -\max(U'(y))$ which generates the same velocities $\pm U_\infty$ at $y = \pm \infty$.

G. Corcos' [4] model of vorticial structures that develop from perturbations of the velocity profile $U(y)$ is based on the following observations: The basic dynamics is two dimensional. The vorticity of the original parallel flow is endlessly redistributed to produce successive generations of vorticial structures whose unfolding we observe with the passage of time.

During the first generation, the vorticity is redistributed into a periodic array of vortex concentrations. The horizontal and vertical concentration of these vortex concentrations are both $O(\delta)$. Figure 4 shows a sequence of snapshots from a numerical simulation by G. Corcos of this first generation rollup. The contours correspond to constant values of the vorticity. The numerical simulations show that the vorticity accumulates into spiral cores which are connected by thin filaments called braids. The dynamics of the rollup is to deplete the vorticity in the braids and redistributed it in the cores. Figure 5 illustrates the physical mechanism involved. The circulatory motions due to adjacent cores create a stagnation point flow in the center of the braid between them which disperses the vorticity from the braid and sends it to the cores.

The array of vorticies produced in the first generation is unstable to subharmonic perturbations. The most prominent instability is the pairwise coalescence of vorticies to produce a new array with twice the wavelength and twice the height. Figure 6 is a sequence of snapshots from a numerical simulation of the pairing process.

The successive generations of vortical structures are due to the continuation of the pairing process. With the passage of each generation, the vorticies double in size and circulation. The time required for pairing of vorticies in a given generation to produce the array of vortices in the next generation is proportional to (vortex size)$^2$/circulation. We see that this time doubles for each successive generation.

A consequence of the hierarchical pairing process is that the height h of vortical structures should be proportional to the time t elapsed since the beginning of the process. The dynamics of the large vortical structures is basically inviscid. The only velocity present in the inviscid problem is $U_\infty$. Hence, we expect

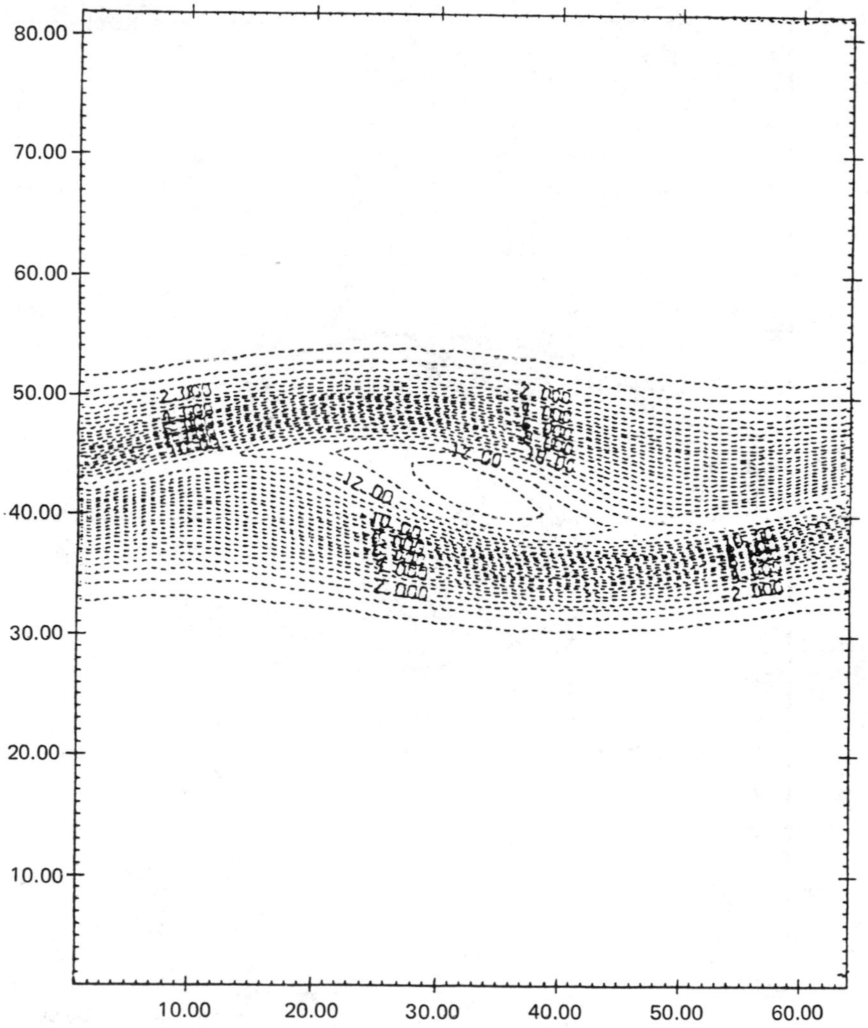
80.00
70.00
60.00
50.00
40.00
30.00
20.00
10.00
10.00
20.00
30.00
40.00
50.00
60.00
12.00
12.00

Figure 4(a)

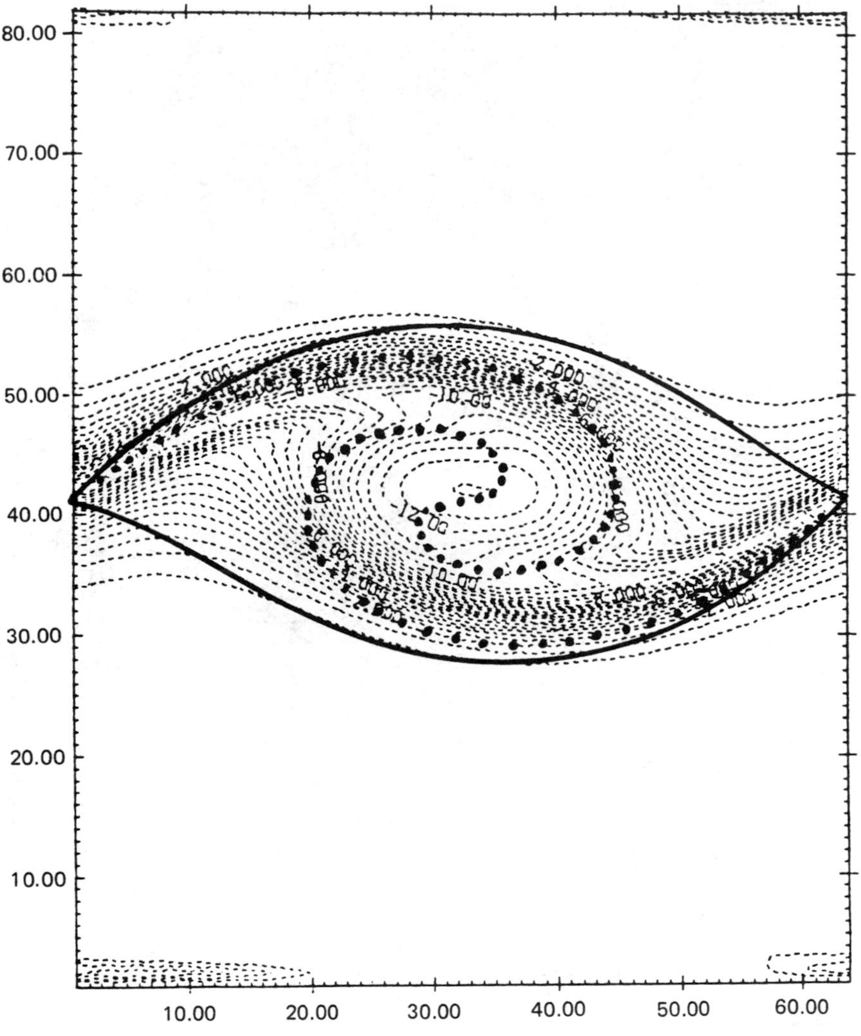
80.00
70.00
60.00
50.00
40.00
30.00
20.00
10.00
10.00
20.00
30.00
40.00
50.00
60.00

Figure 4(b)

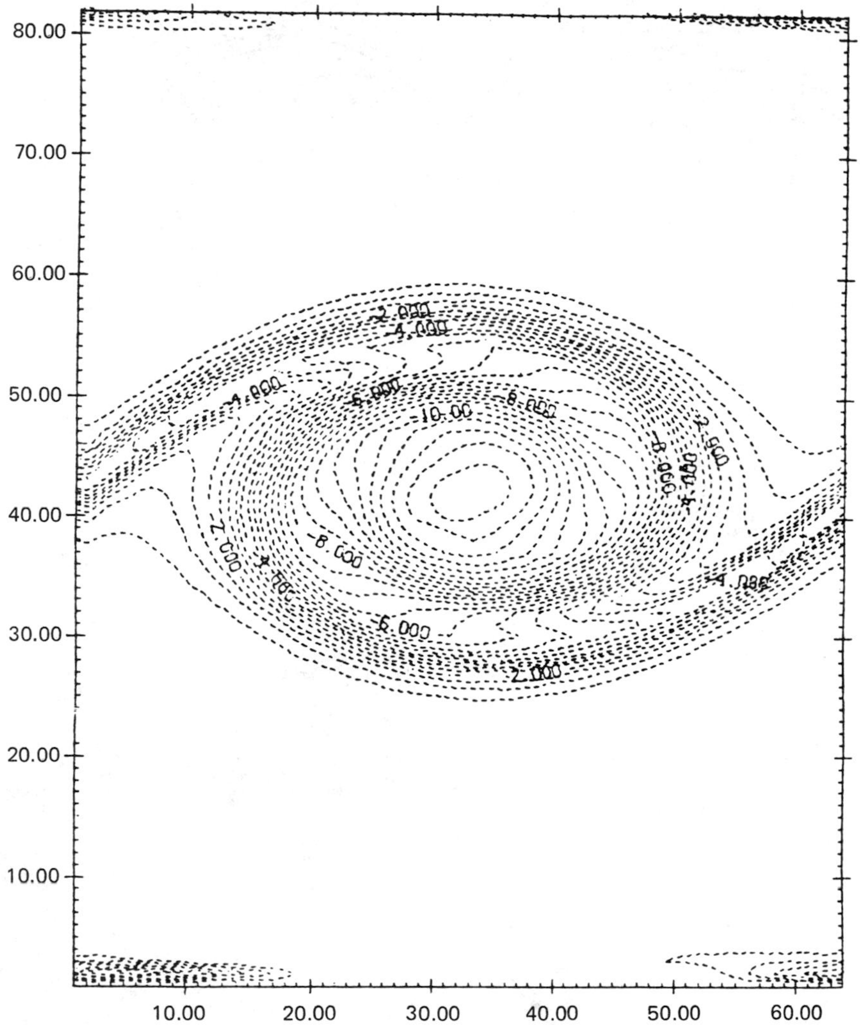
80.00
70.00
60.00
50.00
40.00
30.00
20.00
10.00
10.00
20.00
30.00
40.00
50.00
60.00
-10.00

Figure 4(c)

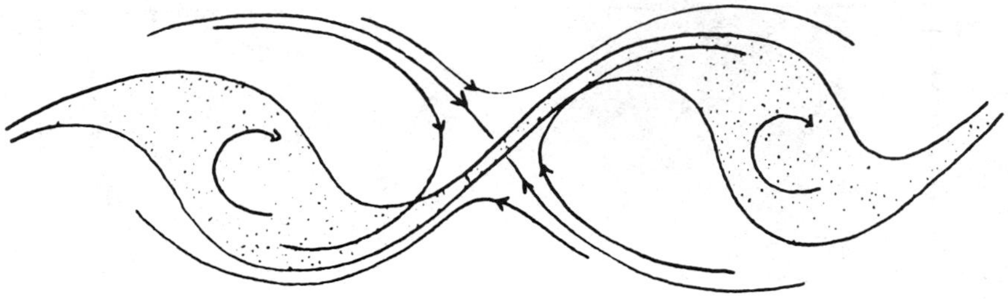

Figure 5

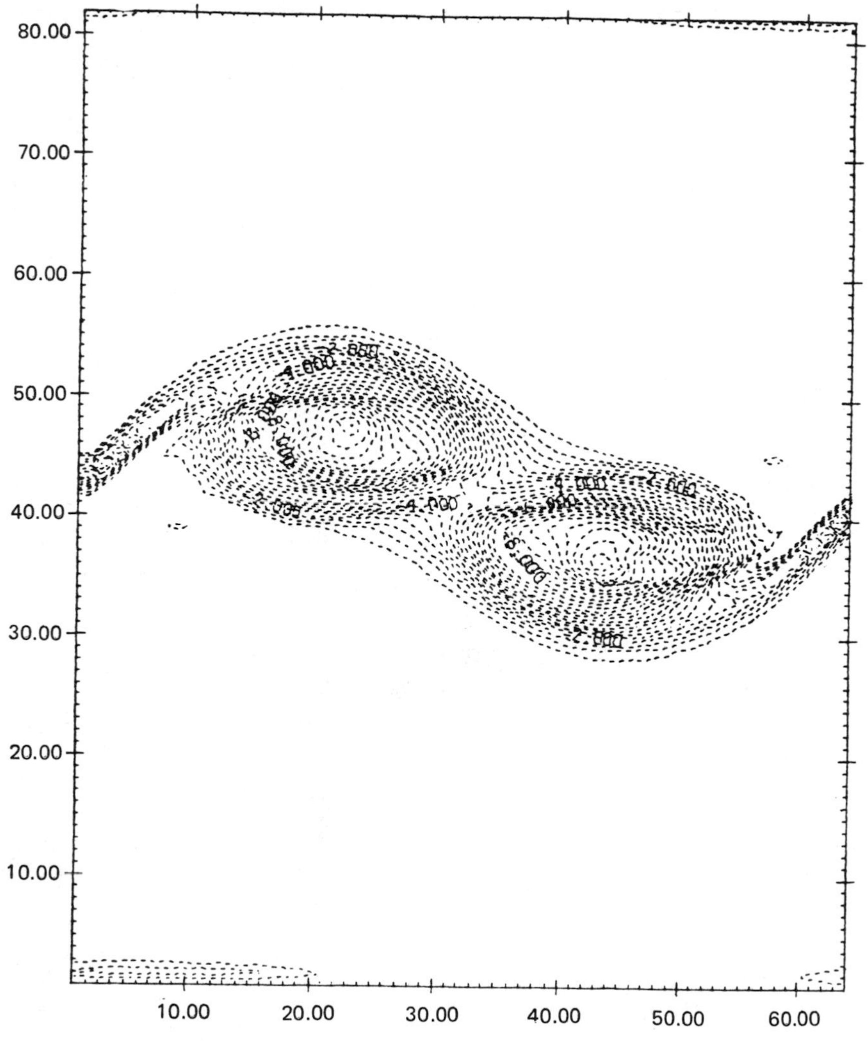
80.00
70.00
60.00
50.00
40.00
30.00
20.00
10.00
10.00
20.00
30.00
40.00
50.00
60.00

Figure 6(a)

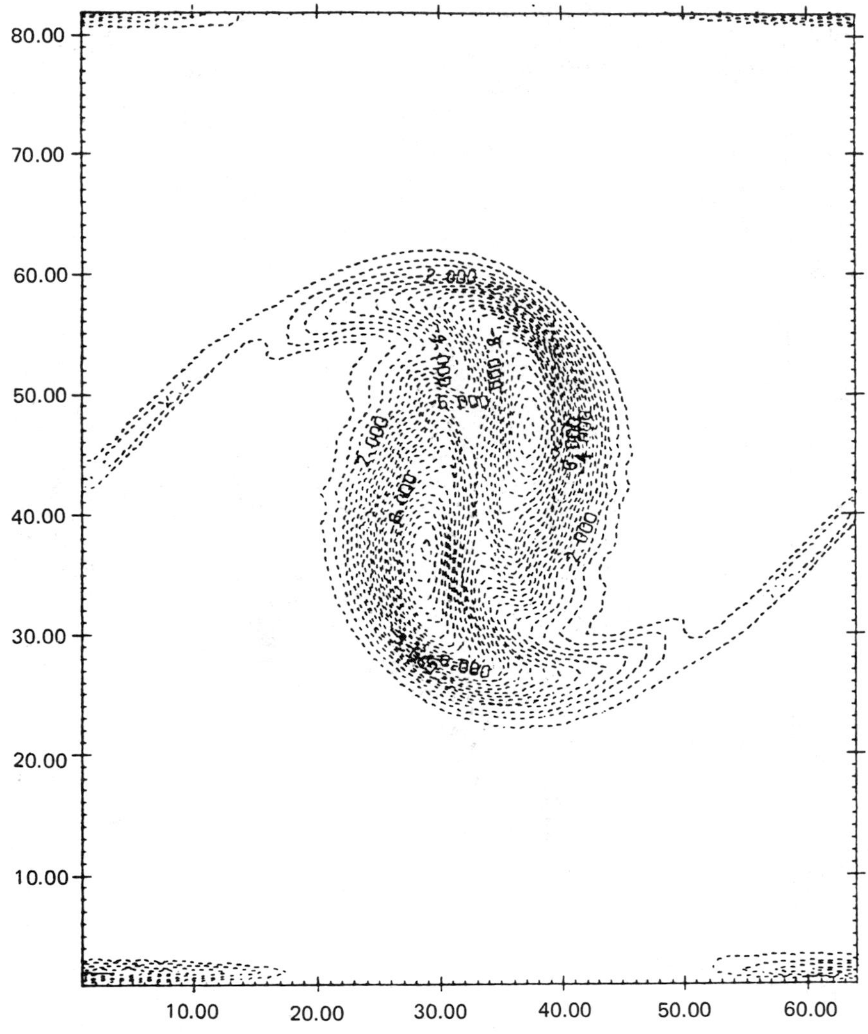
80.00
70.00
60.00
50.00
40.00
30.00
20.00
10.00
10.00
20.00
30.00
40.00
50.00
60.00

Figure 6(b)

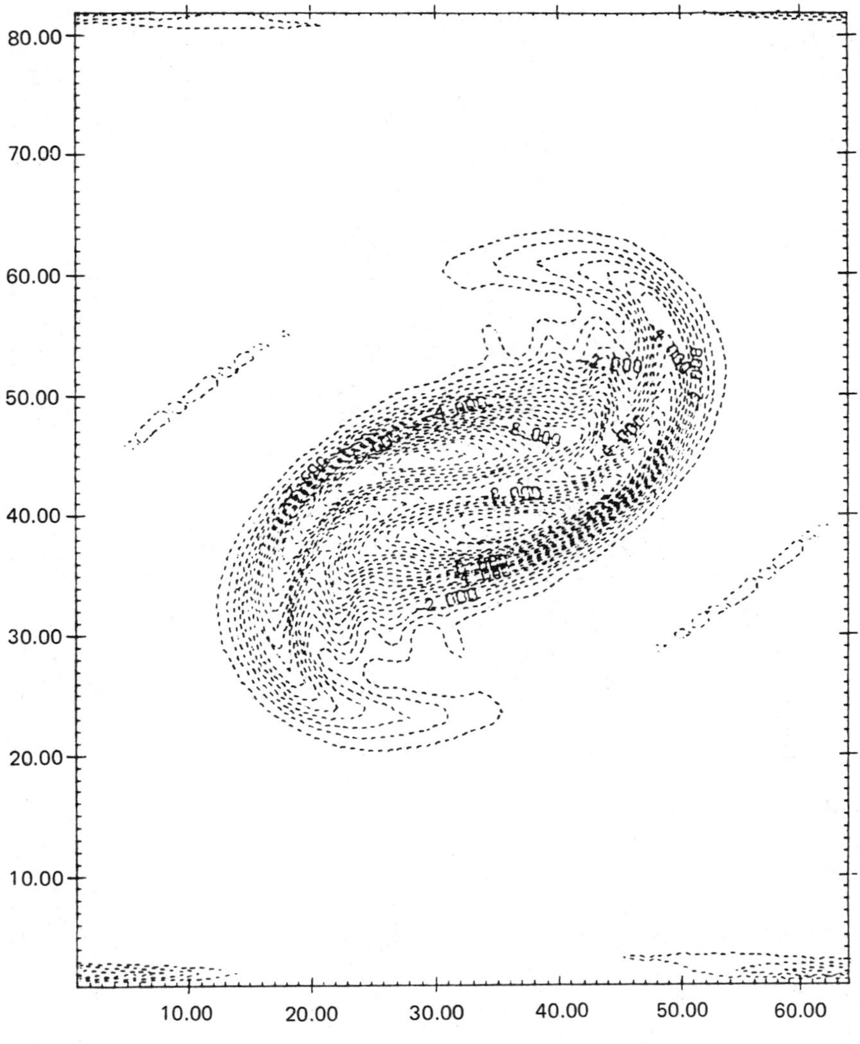
80.00
70.00
60.00
50.00
40.00
30.00
20.00
10.00
10.00
20.00
30.00
40.00
50.00
60.00

Figure 6(c)

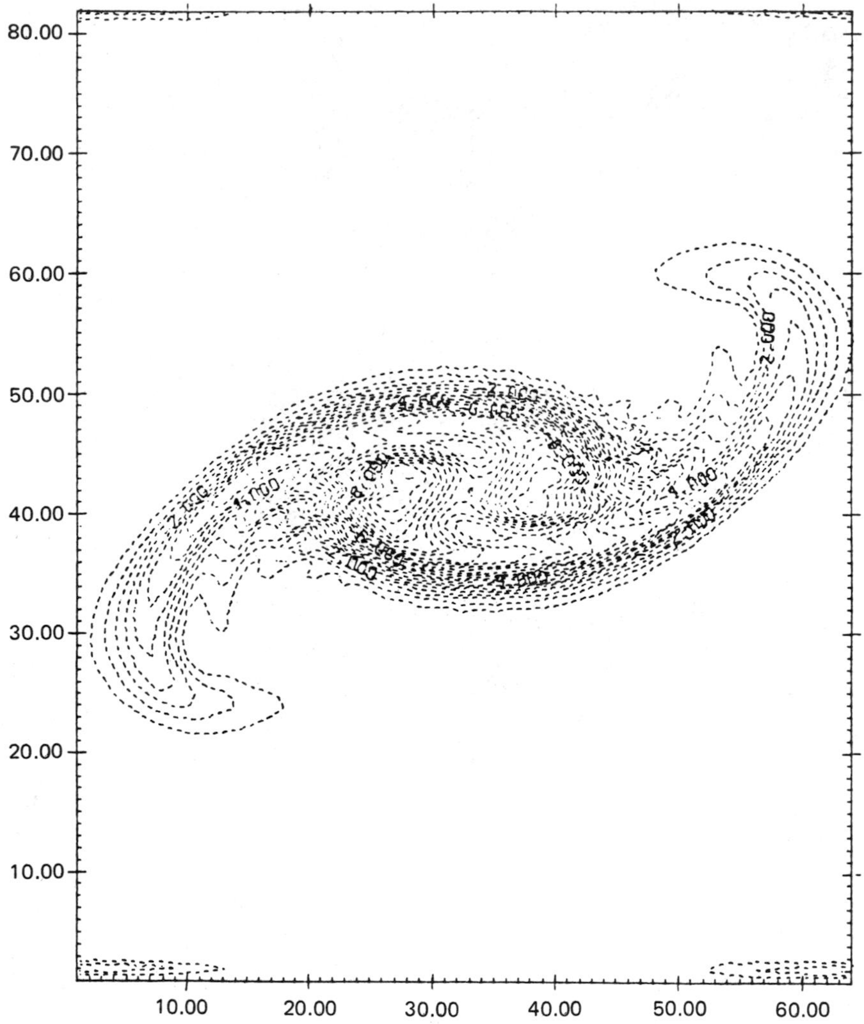
80.00
70.00
60.00
50.00
40.00
30.00
20.00
10.00
10.00
20.00
30.00
40.00
50.00
60.00

Figure 6(d)

$$h = O(U_\infty t). \tag{5.2}$$

We compare this result to the thickness of a vortex layer in which the flow remains parallel, and vorticity in transported only by viscous diffusion. If the initial condition is a vortex sheet with circulation $2U_\infty$ per unit length, then the vorticity $w = w(y,t)$ satisfies the initial value problem

$$w_t = vw_{yy} \text{ in } -\infty<y<\infty \text{ for } t>0, \; w(y,0) = 2U_\infty \delta(y). \tag{5.4}$$

The solution is

$$w = \frac{U_\infty}{(\pi vt)^{1/2}} \exp\left(-\frac{y^2}{4vt}\right). \tag{5.5}$$

The thickness $\delta$ of the associated velocity profile is

$$\delta = (\pi vt)^{1/2}. \tag{5.6}$$

Hence, distribution of vorticity by bulk fluid motion is ultimately much more effective than viscous diffusion.

We now consider a typical experimental situation depicted schematically in figure 7. According to the pairing model, the height of vortex structures at a station moving downstream with the average velocity $\frac{1}{2}(U_1 + U_2)$ should be proportional to $\frac{1}{2}(U_2 - U_1)t$. Hence, the region of turbulent flow at position x downstream from the splitter plate should have thickness

$$h = C\,\frac{U_2 - U_1}{U_1 + U_2}\,t. \tag{5.7}$$

Where C is a constant. R. Brown and A. Roshko [5] find the empirical value C =.38 from their experiments.

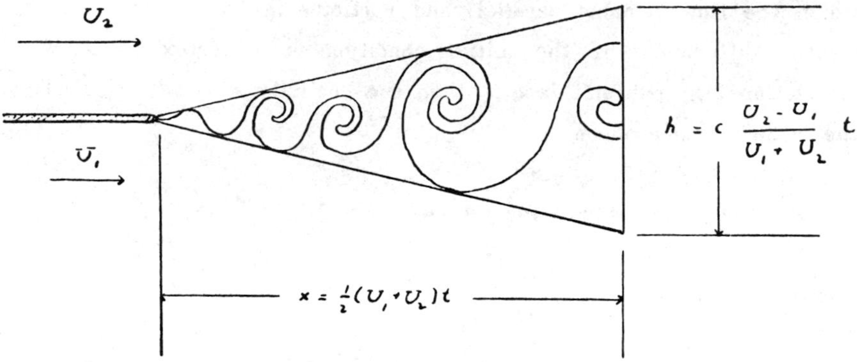


Figure 7

## 6. Secondary vortical structures.

The large two dimensional vortical structures in shear layers provide the background flow for an array of alternating secondary vorticies whose axes follow the braids in the streamwise direction.

At high values of the Reynolds number $Re = \frac{\lambda U_\infty}{2}$ based on the spatial scale $\lambda$ of the primary vorticies, these secondary vorticies appear as highly flattened ribbons. The physical situation is depicted in figure 8.

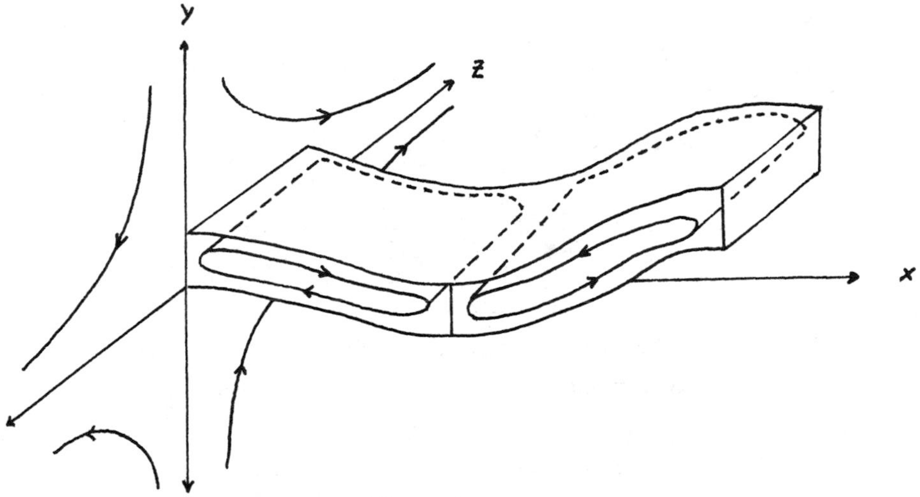


Figure 8

The xz plane coincides with the central portion of a braid between two primary vorticies. The streamlines drawn in the yz plane are those of the stagnation point flow $\underline{u}_e$ created by the primary vorticies. The streamlines of $\underline{u}_e$ are the same in every plane parallel to the yz plane.

S. Lin [6] formulated a simple model of secondary vorticies. In the language of vortex dynamics, the essentials of his model may be described as follows: Near the center of the braid, the stagnation point flow $\underline{u}_e$ has the approximate form

$$\underline{u}_e = -\gamma y\hat{y} + \gamma z\hat{z}, \tag{6.1}$$

where $\gamma$ is a constant called the strain rate. We consider a stage of the evolution in which most of the x vorticity is depleted from the braids and sent to the cores, in which case only the secondary vorticies with their axes in the z direction are present and the vorticity field has the form

$$\underline{w} = w(x,y,t)\hat{z}. \tag{6.2}$$

The vortex dynamics equations (1.3), (1.4) give

$$w_t + uw_x + vw_y - \gamma w = \nu \Delta w, \tag{6.3}$$

where the velocity $u\hat{x} + v\hat{y}$ is given by

$$u(z,t) - v(z,t) = \frac{-i}{2\pi} \int \frac{w(z',t)}{z-z'} \, dx' \, dy' + i\gamma y. \tag{6.4}$$

The integral in (6.4) gives the velocity induced by the vortex distribution. The terms $-\gamma w$ in (6.3) and $i\gamma y$ in (6.4) represent the tendency of the external flow to compress the vortex layer into the xz plane.

S. Lin's observations based on his numerical simulations with this model are these: If the alternating vorticies are sufficiently weak, they rotate slightly until this tendency to rotate is counterbalanced by the restoring effect of the external flow. The opposite vorticities of neighboring vorticies interdiffuse and the array eventually disappears. If the alternating vorticies are sufficiently strong, the destabilizing tendency of convective transport overcomes the stabilizing effect of viscosity and the vorticity focuses into periodic concentrations. The ribbons subsequently buckle and utimately, an array of concentrated circular vorticies results. Figure 9 is a sequence of snapshots from a numerical simulation by S. Lin of the collapse process.

The focusing of the vorticity may be studied asymptotically. We define

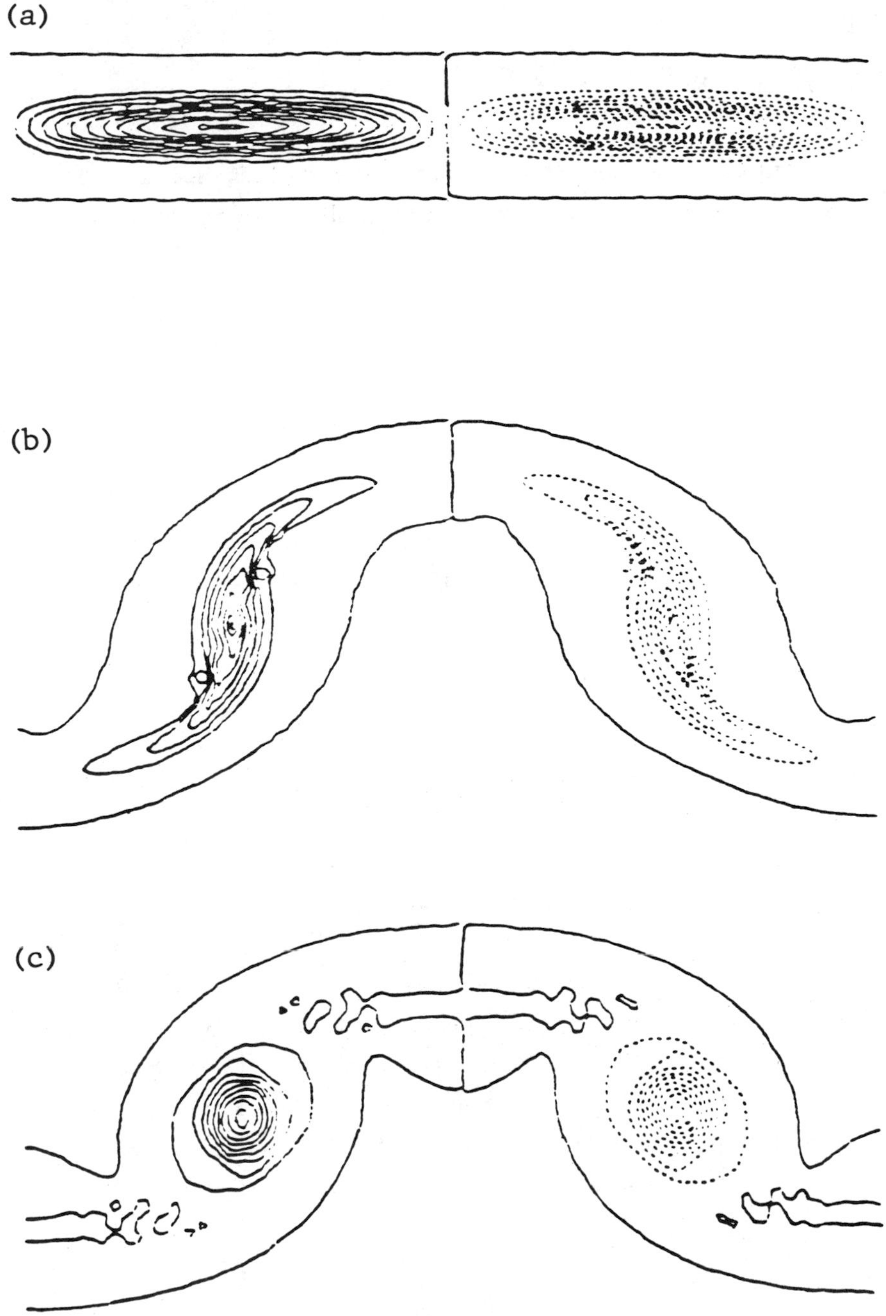

Figure 9

$$\sigma(x,t) \equiv \int_{-\infty}^{\infty} w(x,y,t)\ dy, \tag{6.5}$$

which may be thought of as an effective surface vortex density: $\int_{x_1}^{x_2} \sigma(x,t)dx$ gives the total circulation between $x_1$ and $x_2$. If the gradient of the surface density is much smaller than the strain, that is

$$|\sigma_x| << \gamma \tag{6.6}$$

and

$$\sigma = 0((v\gamma)^{1/2}), \tag{6.7}$$

so that viscous transport balances convective transport, then we may show from the vortex dynamics equations (6.3), (6.4) that $\sigma$ satisfies to leading order [7]

$$\sigma_t = \left[\left[v - \frac{\sigma^2}{4\gamma}\right] \sigma_x \right]_x. \tag{6.8}$$

We recognize (6.8) as a nonlinear heat equation in which the flux F is given by

$$F = v\sigma_x - \frac{\sigma^2}{4\gamma}\sigma_x. \tag{6.9}$$

The first term is the usual viscous transport. The second term represents the convective transport in the limit defined by (6.6), (6.7). It is a negative diffusion, and so represents the focusing effect.

A simple heuristic argument illumines the focusing mechanism. Figure 10 depicts a ribbon of vorticity that has rotated though a small angle until it is balanced by the external straining flow.

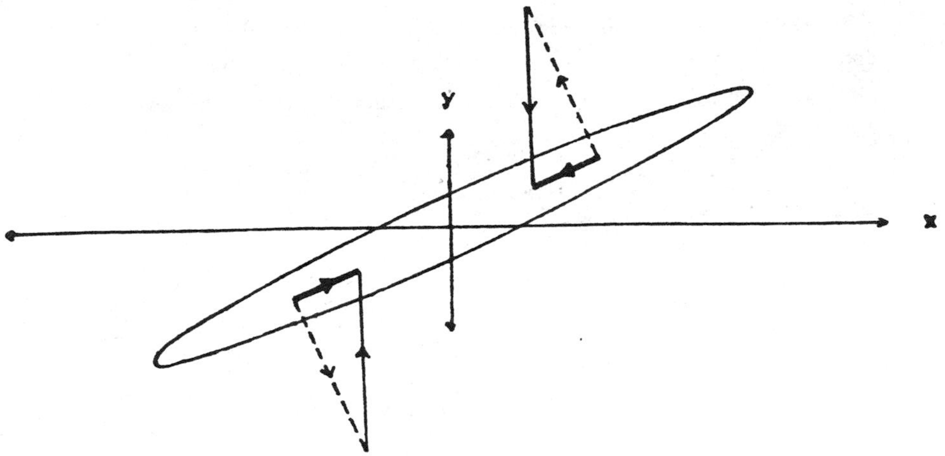


Figure 10

The dashed arrows represent the self induced velocity of the ribbon. The vertical arrows represent the restoring tendency of the external flow. The heavy black arrows represent the vector sums of the previous two. They are directed so as to cause the collapse of the filament.

We see that the convective focusing will predominate over viscous diffusion in (6.8) if

$$|\sigma| > 2\sqrt{\nu\gamma}. \tag{6.10}$$

(6.10) is a criterion for the collapse of the secondary vorticies.

## 7. Vorticies in straining flow.

In many flows, one is interested in the interaction between neighboring vortical structures. The nature of the interaction is that the irrotational far field of each vortex deforms the structure of its neighbors. These far fields are in turn modified by the deformations. Ideally, one would like to describe the full self consistent interaction. In practice, such interaction must be studied numerically.

A related but simpler question is to ask how a given externally imposed, irrotational flow deforms vortex structures. The answer to this question may in some cases by used to model a full self consistent interaction. As an example, consider a pair of counter-rotating vorticies traveling together to the right as shown in figure 11a.

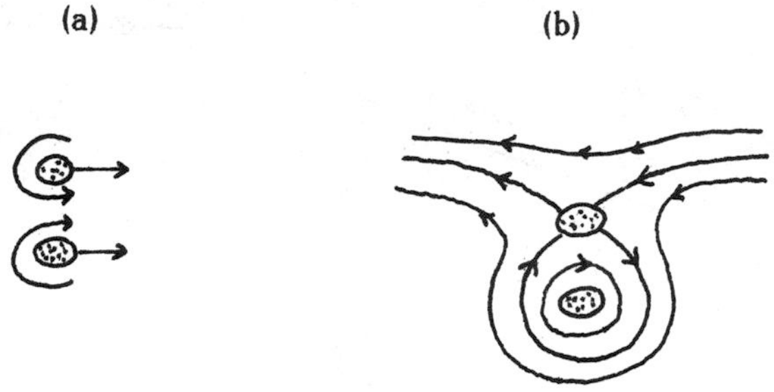


Figure 11

Figure 11b shows the streamlines of the flow due solely to the bottom vortex as seen in a frame of reference traveling with the pair. We see that the bottom vortex produces a straining flow in the vicinity of the top vortex. If the spatial extent of the top vortex is sufficiently small, we may model its structure by assuming that it resides in a uniform straining flow.

There are exact solutions due to D. Moore and P. Saffman [8] representing elliptical vorticies in a uniform straining flow. The boundary of the vortex containing uniform vorticity $w_0$ is given by

$$\frac{x^2}{a^2} + \frac{y^2}{b^2} = 1. \tag{7.1}$$

The far field flow is

$$u \sim -\epsilon y + \gamma y, \; v \sim -\epsilon x - \gamma y \text{ as } x^2 + y^2 \longrightarrow \infty, \tag{7.2}$$

which is a superposition of a uniform straining flow with strain rate $\epsilon$ and a solid body rotation with angular velocity $\gamma$. In order for there to be a flow with the above specifications, the parameters $\epsilon$,

$\gamma$, $w_0$ and $\theta \equiv a/b$ must be related by

$$\frac{\epsilon(\theta + 1)}{\theta - 1} - \gamma = w_0 \frac{\theta}{\theta^2 + 1}. \tag{7.3}$$

For various values of $\epsilon$, $\gamma$, $w_0$, the number of solutions for $\theta$ varies between 0 and 3.

In the special case of a pure straining flow with $\gamma = 0$, (7.3) has two solutions for $\theta$ if $\epsilon/w_0 < .15$, and no solution for $\epsilon/w_0 > .15$. Of the two possible vorticies for $\epsilon/w_0 < .15$, only the least elongated is stable to two dimensional perturbations. These results suggest that a vortex becomes irreversibly elongated if it is subjected to a strain that is sufficiently large in comparison to its vorticity.

Moore and Saffman's solution treats the <u>statics</u> of elliptical vorticies in a strain field. It is also possible to study the time dependent deformations of elliptical vorticies in a strain field. This possiblity owes itself to the following circumstances: The self induced velocity inside of an elliptical patch of uniform vorticity is linear in $\underline{x}$:

$$\underline{u}_s = V(n,\theta)\underline{x}, \tag{7.4}$$

$$V(n,\theta) = -\frac{w_0}{1+n} R(\theta) \begin{bmatrix} 0 & n \\ -1 & 0 \end{bmatrix} r(-\theta), \tag{7.5}$$

where $\theta$ is the angle of orientation of the ellipsis' major axis. $R(\theta)$ is the matrix for rotation through angle $\theta$ in the plane, and n is the ratio of the major and minor axes. If we superimpose upon $\underline{u}_s$ the strain induced velocity

$$\underline{u}_e = -\gamma \begin{bmatrix} 0 & 1 \\ 1 & 0 \end{bmatrix} \underline{x}, \tag{7.6}$$

We obtain the total velocity

$$\underline{u} = \left[ V(n,\theta) - \gamma \begin{bmatrix} 0 & 1 \\ 1 & 0 \end{bmatrix} \right] \underline{x}. \tag{7.7}$$

Hence, the sum of self induced and strain induced velocities is still linear in $\underline{x}$. The flow of points in the plane induced by (7.7) is area preserving, since $V - \gamma \begin{bmatrix} 0 & 1 \\ 1 & 0 \end{bmatrix}$ is traceless. It follows that an elliptical patch of uniform vorticity will always remain an ellipse with the same area. The dynamics of the flow is fully specified if we know the evolution of the ellipse parameters n and $\theta$. These equations are

$$\dot{n} = -2\gamma \sin 2\theta n, \tag{7.8}$$

$$\dot{\theta} = w_0 \frac{n}{(1+n)^2} - \gamma \frac{n^2+1}{n^2-1} \cos 2\theta. \tag{7.9}$$

This system has an integral

$$H = w_0 \log \frac{(1+n)^2}{n} - \gamma(n - \frac{1}{n})\cos 2\theta. \tag{7.10}$$

Figure 12 is a sequence of phase portraits of the system (7.8), (7.9) for various values of $\epsilon/w_0$. These phase portraits show the possible trajectories in the plane whose polar coordinates are $(n,\theta)$.

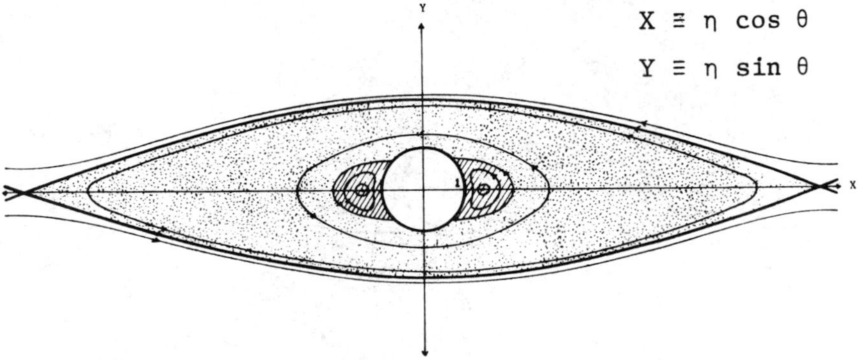


$X \equiv \eta \cos \theta$

$Y \equiv \eta \sin \theta$

Figure 12(a) $\varepsilon/\omega_0 = .084$

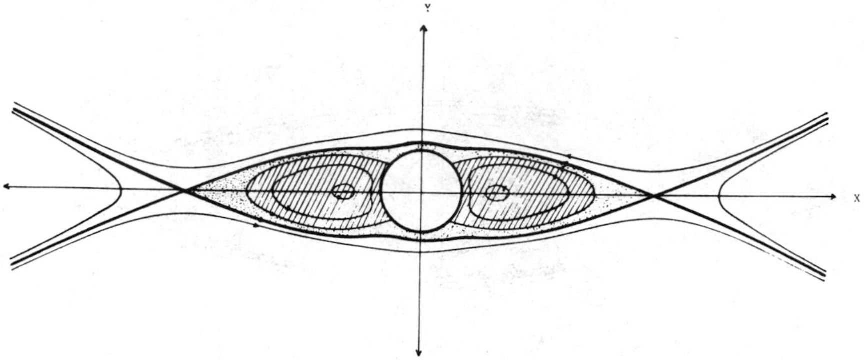


Figure 12(b) $\varepsilon/\omega_0 = .120$

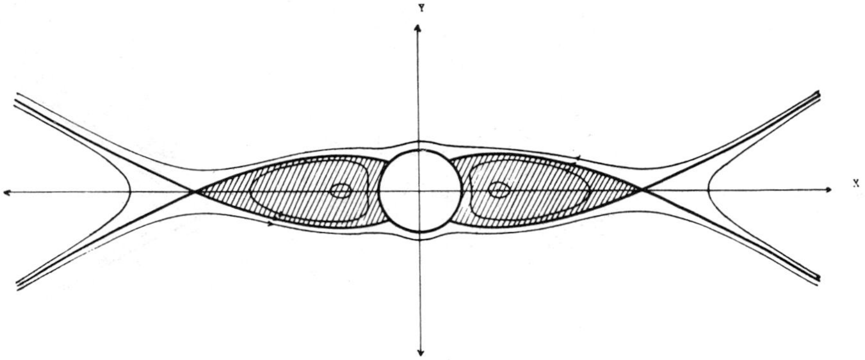


Figure 12(c) $\varepsilon/\omega_0 = .1227$

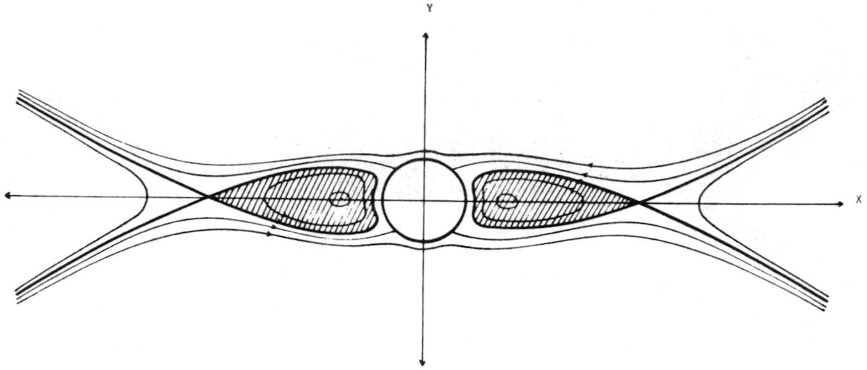


Figure 12(d) $\varepsilon/\omega_0 = .126$

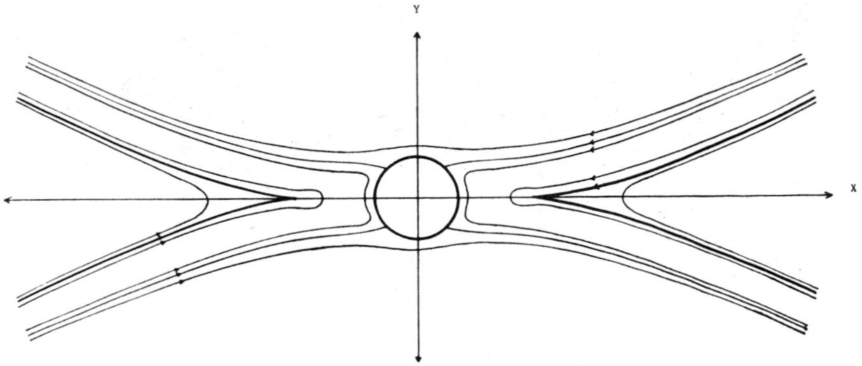


Figure 12(e) $\varepsilon/\omega_0 = .15$

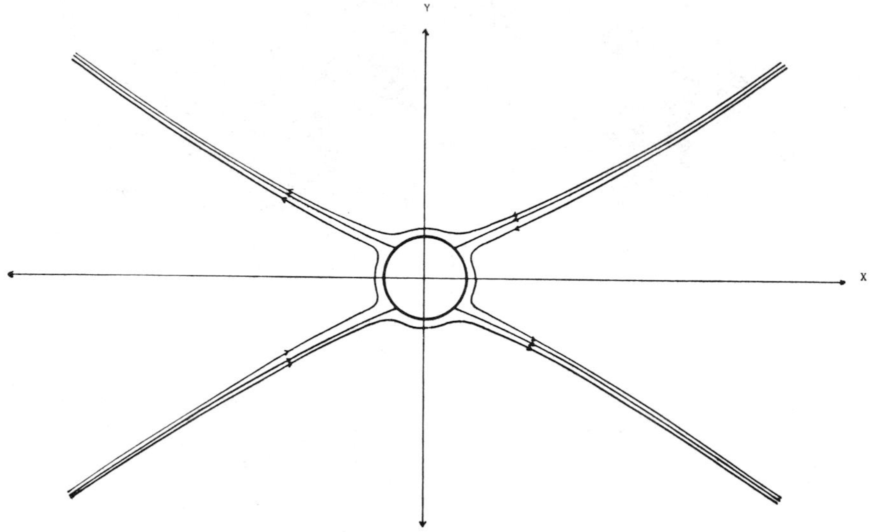


Figure 12(f) $\varepsilon/\omega_0 = .25$

## References

[1] G. Birchoff, Proc. Sym. Appl. Math., Am. Math. Soc. 13.

[2] D. Moore, Proc. Roy. Soc. 365A.

[3] D. Merion, G. Baker, S. Orzag, JFM Vol. 114.

[4] G. Corcas, "The mixing layer, deterministic models of turbulent flow," College of Engineering Report (1979).

[5] R. Brown, A. Roshko, "Structure of turbulent shear flows: A new look." AISA Reprint 76-78 (1976).

[6] S. Lin, "The evolution of streamwise vorticity in the ??????? laws," College of Engineering Report (1981).

[7] J. Neu, "The dynamics of stretched vorticies," To appear in JFM.

[8] D. Moore, P. Saffman, "Vortex interaction," Ann. Rev. Fluid Mech. 11.

# THE MINIMAL SURFACE EQUATION

R. Osserman

The <u>minimal</u> <u>surface</u> <u>equation</u> is a system of non-linear elliptic partial differential equations of the form

$$(1) \qquad \sum_{i,j=1}^{n} a_{ij} \frac{\partial^2 y_k}{\partial x_i \partial x_j} = 0 , \qquad k = 1,\ldots,m ,$$

where the coefficients $a_{ij}$ are certain functions of the first-order derivatives $\partial y_r/\partial x_s$, and the matrix $(a_{ij})$ is positive definite for all values of the arguments. It is an unusual system in that, on the one hand it is made simpler by the fact that it is very weakly coupled: a solution is a set of functions

$$(2) \qquad y_k = f_k(x_1,\ldots,x_n) , \qquad k = 1,\ldots,m ,$$

all satisfying the <u>same</u> equation, with the coupling only appearing in the coefficients $a_{ij}$, which depend on all the $f_k$. On the other hand, the form of the coefficients is such that even when presented with a specific set of elementary functions $f_k$, it may be difficult to verify that they constitute a solution of (1). The reason is that the $a_{ij}$ are defined in terms of the entries in an inverse matrix. Specifically, if $G = (g_{ij})$ is the matrix defined by

$$(3) \qquad g_{ij} = \delta_{ij} + \sum_{k=1}^{m} \frac{\partial f_k}{\partial x_i} \frac{\partial f_k}{\partial x_j} , \qquad i,j = 1,\ldots,n ,$$

then the $a_{ij}$ are the elements of

(4) $A = (\det G)G^{-1}.$

It follows from (3) and Cramer's rule, that the coefficients $a_{ij}$ are polynomials of degree at most $2(n-1)$ in the partials $\partial f_k/\partial x_i$.

As an example of the above remarks, it is proved by general arguments in [21] that the functions

$$f_1(x) = \sqrt{\tfrac{5}{2}}\,\tfrac{1}{r}(x_1^2 + x_2^2 - x_3^2 - x_4^2)$$

$$f_2(x) = \sqrt{\tfrac{5}{2}}\,\tfrac{1}{r}(x_1x_3 + x_2x_4) \tag{5}$$

$$f_3(x) = \sqrt{\tfrac{5}{2}}\,\tfrac{1}{r}(x_2x_3 - x_1x_4)$$

where $r = |x|$, satisfy (1) for all $x \neq 0$ in $\mathbb{R}^4$. However, it would be a sizeable task to compute the coefficients $a_{ij}$ by substituting (5) in (3) and (4), and then to verify directly that the functions (5) satisfy equation (1).

Given the complexities and the degree of non-linearity of this system of equations, it is not surprising that except in the case $m = 1$, when the system reduces to a single equation, general P.D.E. methods are of little use. Basically, all results in the general case depend on the same fact that provides the impetus for studying the system in the first place — that the equations (1), (3), (4) are the necessary and sufficient conditions for the submanifold (2) in $\mathbb{R}^{m+n}$ defined by the graph of the functions $f_k$, to be a <u>minimal submanifold</u>, in the sense that its mean curvature vector is zero. That fact, incidentally, although not hard to prove, is not quite obvious either, since it requires showing that every solution of the system (1) must also satisfy the additional equations

$$\sum_{i=1}^{n} \frac{\partial}{\partial x_i}(\sqrt{g}\, g^{ij}) = 0\ , \quad j = 1,\dots,n\ , \tag{6}$$

where the $g^{ij}$ are the entries of the matrix $G^{-1}$, and $g = \det G$. (For details, see [37], Th. 2.2.) Equations (6) play a critical role in the theory for the case $n = 2$ (see [36], §3, and [38]) but to my knowledge they have never been used in the general case.

Another way to think of equation (1) (or equivalently, (1) and (6) combined) is as the Euler-Lagrange equation for the area integrand of a submanifold represented as a graph; that is, equation (1) expresses the property that the first variation of area vanishes for all deformations of the surface leaving the boundary fixed.

There is a long history of attempts to study the minimal surface equation, and as often as not the methods introduced have proved of much wider impact than the specific problem addressed. For example, Lebesgue's invention of the Lebesgue integral in his thesis [22] is related to his investigation of area and its application to the solution of the Dirichlet problem for the minimal surface equation.

Of the considerable literature devoted to solutions of the minimal surface equation, the part that has undoubtedly excited most interest is that related to Bernstein's Theorem [2]: <u>when</u> $n = 2$, $m = 1$, <u>the only entire solutions of the minimal surface equation are linear functions</u>. (An entire solution is one defined for all $(x_1,\ldots,x_n)$.) Bernstein's theorem turns out to be closely related to another topic of great interest: the possible singularities of minimal submanifolds. A third major area of study is the Dirichlet problem (which is in turn related to the Plateau problem). Since this last area has been the subject of several fairly recent surveys [30], [40], we shall forego any detailed discussion, and concentrate on the subject of entire solutions. In fact, many of the seminal ideas for the whole theory have originated in attempts to settle the basic questions regarding Bernstein's theorem and its generalizations.

To provide a convenient framework for presenting the known facts, I have selected eight of what seem to me to be among the most seminal papers in terms of future developments. For the most part I have omitted discussion of the classical case of one equation in two unknowns, both because it is thoroughly covered in Nitsche's book [33], and because for the most part the methods have not generalized. For further details in our general subject area, there are two excellent recent surveys by Nitsche [34] and Yau [52], as well as the book by Gilbarg and Trudinger [11].

## 1. The fountainhead [2].

Bernstein's theorem for minimal surfaces originally appears as a kind of afterthought. It is presented as an application of a deep result known as "Bernstein's geometric theorem", stating that a function $f(x,y)$ whose graph has Gauss curvature $K \leq 0$ in the whole $x,y$-plane, and $K < 0$ at some point, cannot be bounded. As an application of that theorem, Bernstein derives a very general Liouville Theorem to the effect that any bounded entire solution to a differential equation

$$(7) \qquad Au_{xx} + 2Bu_{xy} + Cu_{yy} = 0 \ , \quad AC - B^2 > 0 \ ,$$

must be constant. The coefficients $A, B, C$ may be arbitrary functions of $x$ and $y$ as well as $u$ and its derivatives.

An immediate consequence is that every bounded entire solution of the minimal surface equation is constant. However Bernstein observes that if $f(x,y)$ satisfies the minimal surface equation, then the functions

$$u = \tan^{-1} f_x \quad \text{and} \quad v = \tan^{-1} f_y$$

satisfy the equation

$$(1 + f_y^2)u_{xx} - 2f_x f_y u_{xy} + (1 + f_x^2)u_{yy} = 0 \ ,$$

which is of the form (7). It follows that $f_x$ and $f_y$ are both constant. Thus Bernstein derives the far stronger and much more unexpected result that <u>any</u> entire solution (with no boundedness or growth assumptions) must be linear.

The only flaw in Bernstein's argument was a gap in his proof of the Geometric Theorem. It was not until 1950 that a complete proof was given [19], [27]. In the meantime, a new proof of Bernstein's minimal surface theorem was given by Radó [42] using complex function theory. A number of other complex analysis proofs have since been obtained, most notably those of Bers [3] and Nitsche [32].

## 2. The Heinz inequality [17].

Noting that the auxiliary functions u, v introduced by Bernstein are harmonic with respect to any isothermal parameters on the surface, Erhard Heinz obtained an important sharpening of Bernstein's Theorem in the form of a finite version. For any solution f(x,y) of the minimal surface equation in a disk of radius R, the second derivatives of f at the center of the disk must be bounded by expressions in the first derivatives at the center; for example

$$(8) \qquad f_{xx}^2 + 2f_{xy}^2 + f_{yy}^2 \leqslant \frac{2}{3}\pi^2 \frac{(2 + f_x^2 + f_y^2)^3}{R^2} .$$

For an entire solution, we may choose a disk about any point, and letting $R \longrightarrow \infty$, we conclude that all second derivatives are zero. Thus Bernstein's theorem appears as a limiting case. It was then observed by Heinz and Rellich ([17], p.55) that one could replace (8) by the more geometric statement

$$(9) \qquad |K| \leqslant \frac{c}{R^2} ,$$

where c is a universal constant, and K is the Gauss curvature of the surface $z = f(x,y)$ at the center of the disk. Heinz's inequality (9) was improved by various authors, including Hopf [20] and Osserman [35]. There has also been much work on determining the best value of the constant $c$, but that remains an open problem.

## 3. Fleming's proof [10] using GMT.

The most important single paper for future developments was one of Fleming where the newly developed tools of geometric measure theory were substituted for the methods of PDE and function theory. Although the main concern of the paper is to give a solution of Plateau's problem for oriented surfaces, Fleming observes that basically the same technique used to show that the surface obtained has no singularities also provides an entirely new proof of Bernstein's Theorem. Namely, to demonstrate the regularity of the surface at a point $p$, a sequence of surfaces is constructed in the unit ball centered at $p$ by successive magnifications of the original surface. One is led in this way to study minimal cones with vertex at $p$. In the case of an entire solution of the minimal surface equation, given any point $p$ on the corresponding surface one can form a sequence of contractions, and again obtain a cone in the unit ball about $p$. In this way, Bernstein's theorem is reduced to proving that if $C$ is a cone with vertex at the origin and boundary on a sphere centered at the origin such that $C$ has least area among all surfaces with the same boundary, then $C$ is a Euclidean disk. For two-dimensional surfaces in $\mathbb{R}^3$ that fact is easily established, and yields a new proof of Bernstein's theorem.

The fundamental importance of Fleming's paper is that his argument works for minimal hypersurfaces in any number of dimensions. It therefore provided the first real hope of proving the general "Bernstein Theorem", that an entire solution of the minimal surface equation in any number of variables (still with codimension 1) must be linear. Namely, it sufficed to prove the italicized statement above for minimal cones of codimension 1 in any $\mathbb{R}^n$.

The first step in that direction was taken by Almgren [1] who showed the result was true for 3-dimensional cones in $\mathbb{R}^4$. Even before that, deGiorgi [9] had found an improvement in Fleming's argument; he showed that the statement about minimal k-cones in $\mathbb{R}^{k+1}$ implied Bernstein's Theorem for minimal graphs over $\mathbb{R}^{k+1}$ in $\mathbb{R}^{k+2}$. This gave Bernstein's Theorem for 3 variables in $\mathbb{R}^4$ and using Almgren, for 4 variables in $\mathbb{R}^5$. It was beginning to feel like Bieberbach's conjecture, with a new argument for each dimension.

## 4. Simons's breakthrough [50].

Motivated by the desire to extend Bernstein's Theorem to even higher dimensions, James Simons wrote a paper [50] that more than any other was responsible for initiating a new era in the study of minimal surfaces: the detailed theory of minimal submanifolds in spheres and in general Riemannian manifolds. The paper is a goldmine of ideas and results related peripherally at most to the original problem. The two key contributions to the Bernstein problem have to do with the stability of minimal cones. Note that if a surface minimizes area with respect to all surfaces having the same boundary, then for any variation of the surface, keeping the boundary fixed, the first variation of area must be zero, and the second variation cannot be negative. If the second variation is negative, the surface is called unstable. Simons proved first that every k-dimensional cone C in $\mathbb{R}^{k+1}$ is unstable for $k \leqslant 6$, unless C is a k-dimensional plane. Using deGiorgi's modification of Fleming's argument, it follows that Bernstein's Theorem is true for entire solutions of the minimal surface equation in k variables, for $k \leqslant 7$. (It also follows, by the type of argument indicated above, that area minimizing hypersurfaces in $\mathbb{R}^n$ have no singularities when $n \leqslant 7$.)

All hopes of using the same method in higher dimensions were dashed by Simons's other key result; <u>the cone in $\mathbb{R}^8$ defined by</u>

$$(10) \qquad x_1^2 + x_2^2 + x_3^2 + x_4^2 = x_5^2 + x_6^2 + x_7^2 + x_8^2$$

<u>is stable</u>: <u>all variations of the part inside the unit ball, fixing the boundary, are initially area-increasing</u>.

## 5. The bombshell [4].

It was not long after Simons's paper appeared that Bombieri, deGiorgi and Giusti provided the surprise ending to the Bernstein saga. They proved that

1. <u>The "Simons cone" (10) is not only stable but absolutely area-minimizing with respect to its boundary</u>.

In particular, interior regularity of area-minimizing hypersurfaces, true in $\mathbb{R}^n$, for $n \leq 7$, is no longer valid when $n \geq 8$, a point most graphically demonstrated by the vertex of the Simons cone.

2. <u>Bernstein's Theorem is false, in general</u>.

Specifically, for every $k \geq 8$, there exist non-linear entire solutions of the minimal surface equation in $k$ variables.

This startling discontinuity in behavior as a function of dimension is probably unique in all of partial differential equations. Some links with the sudden appearance of exotic spheres in dimension 7 have been suggested, but the supporting evidence seems more numerological than mathematical.

On the other hand, as a discontinuity in the study of Bernstein's Theorem, the effect of this paper was fairly typical. While marking the end of one era, it ushered in the next. Now the obvious questions became:

1. If entire solutions need not be linear, do they have any other distinguishing characteristics?

2. What additional restrictions on entire solutions would guarantee linearity in all dimensions?

Thus, an earlier paper of Moser [31] took on new significance. In it he showed that an entire solution with bounded gradient must be linear. That result was superseded by Bernstein's Theorem for $k \leqslant 7$, but was now of interest for $k \geqslant 8$.

A word may be in order on the construction of the counterexample to the higher-dimensional Bernstein Theorem. Using the great symmetry of the Simons cone, one is led to introduce new variables

$$u = (x_1^2 + x_2^2 + x_3^2 + x_4^2)^{1/2}, \quad v = (x_5^2 + x_6^2 + x_7^2 + x_8^2)^{1/2}.$$

A function $f(x_1,\ldots,x_8)$ of the special form $F(u,v)$ will satisfy the minimal surface equation (1) if and only if $F$ satisfies the equation

$$(1+F_v^2)F_{uu} - 2F_uF_vF_{uv} + (1+F_u^2)F_{vv} + 3\left(\frac{F_u}{u} + \frac{F_v}{v}\right)(1+F_u^2+F_v^2) = 0.$$

The delicate part is to find subsolutions and supersolutions of this equation with greater than linear growth at infinity. One can then use fairly standard techniques to squeeze a minimal surface in between.

## 6. Ruh's Theorem [43].

A clearcut case of the means justifying the end is provided by a paper of Ruh [43]. The goal of the paper is to analyze the global behavior of entire solutions to the minimal surface equation. The result is a sharpening of Moser's Theorem [31] that the gradient cannot be bounded. It makes precise a statement to the effect that "the average gradient" tends to infinity.

Much more important than this result, however, was an auxiliary theorem used in its proof:

*For a minimal hypersurface in $\mathbb{R}^n$ (or more generally, a hypersurface of constant mean curvature), the Gauss map is a harmonic map.*

Specifically, if $M$ is a hypersurface of constant mean curvature, and if $g$ is the Gauss map of $M$ into the unit sphere $S$, Ruh shows that the first variation of energy is zero, for any smooth one-parameter family of maps of $M$ into $S$, starting with $g$. Although with the wisdom of hindsight, such a property seems perfectly natural, bordering on the obvious, the fact is that at the time it represented a significant departure from previous uses of variational methods in geometry. Typically, one varied the manifold $M$, and considered the behavior of various functionals (such as area, energy, etc.) In this case one *fixes* $M$ and considers variations of $g$ through maps having no geometric significance whatever.

Ruh's Theorem, and its later extension to arbitrary codimension by Ruh and Vilms [44], provided a new indispensible tool for studying minimal submanifolds, as well as those of constant mean curvature. The Ruh-Vilms Theorem states that *if $M$ is a submanifold of arbitrary dimension and codimension in $\mathbb{R}^n$, then the generalized Gauss map of $M$ into the Grassmannian is a harmonic map if and only if the mean curvature vector of $M$ is parallel in the normal bundle. In particular, the generalized Gauss map is harmonic for minimal submanifolds of $\mathbb{R}^n$ of any dimension and codimension.*

We should note that for a two dimensional submanifold $M$ in

$\mathbb{R}^n$, the corresponding Grassmannian has a complex structure, and by a theorem of Chern [7], M is minimal if and only if the generalized Gauss map is anti-holomorphic (a much stronger condition than harmonicity).

Returning to Ruh's original goal, a sharpened form of Moser's Theorem, we note that Bombieri and Giusti [6] extended Moser's Theorem in a different direction. They show that if u is an entire solution to the minimal surface equation, and if all but one of the partial derivatives of u are bounded, then u is linear. They also state that combining their methods with those of Simons, one can prove a theorem that contains as special cases both the generalized Bernstein Theorem and Moser's Theorem. Namely, if u is an entire solution of the minimal surface equation, and if all but seven of the partial derivatives of u are bounded, then u is linear.

These and other results in the same paper were motivated by an attempt to prove the conjecture: entire solutions to the minimal surface equation are of polynomial growth. That conjecture is probably the leading candidate to replace the lost linearity in the original Bernstein Theorem. The first step in that direction has apparently just been taken by Leon Simon who proved polynomial growth for entire solutions in eight variables.

## 7. Schoen-Simon-Yau.

A whole new approach to the Bernstein Theorem and related problems was laid out in a paper of R. Schoen, Leon Simon and S.-T. Yau [46]. Among the major innovations of this paper were

1. the first relatively direct proof of the higher-dimensional Bernstein Theorem, not depending on Fleming's detour via limiting currents and cones,

2. the first "finite version", generalizing the Heinz inequality (9),

3. a more general setting, with many of the preliminary results valid for stable minimal hypersurfaces in Riemannian manifolds.

The generalized Heinz inequality takes the form:

<u>There is an absolute constant</u> $\beta$ <u>such that if</u> $u(x_1,\ldots,x_n)$ <u>is a solution of the minimal surface equation in a ball of radius</u> R, <u>and if</u> $\kappa_1,\ldots,\kappa_n$ <u>are the principal curvatures of the hypersurface</u> $x_{n+1} = u(x_1,\ldots,x_n)$ <u>at the point projecting on the center of the ball, then for</u> $n \leqslant 5$ ,

$$(11) \qquad \sum_{i=1}^{n} \kappa_i^2 \leqslant \beta/R^2.$$

Clearly, letting $R \longrightarrow \infty$ gives Bernstein's Theorem for up to 5 variables. (A later paper by Simon [47] derives by a more direct argument the Heinz-type inequality (11) for all $n \leqslant 7$.)

Again, even more important than the result obtained is the injection of new methods. Among the many subsequent applications, we mention the existence and regularity of minimal hypersurfaces in compact manifolds proved by Pitts [41] and improved by Schoen-Simon [45].

### 8. Hildebrandt, Jost and Widman.

All the results discussed so far have been restricted to solutions of the minimal surface equation consisting of a single function. (That is, to the case $m = 1$ in equation (1).) The first

Bernstein-type theorem valid for arbitrary dimension and codimension is due to Hildebrandt, Jost and Widman [18]. They derived a result that in the case of codimension one reduces to Moser's Theorem:

Let (2) be an entire solution of the minimal surface equation (1), (3), (4). Set

$$p = \min\{n,m\}\ , \qquad \kappa = \begin{cases} 1\ , & \text{if} \quad m = 1 \\ 2\ , & \text{if} \quad m > 1\ . \end{cases}$$

If there is a constant $\beta_0$ satisfying

(12) $\beta_0 < 1/\cos^p(\pi/2\sqrt{\kappa p})$

such that

(13) $\sqrt{\det(g_{ij})} \leqslant \beta_0$

everywhere, then each $f_k$ is a linear function of $x_1,\ldots,x_n$.

The proof of this theorem uses the result of Ruh-Vilms [44] referred to earlier; since the graph of the solution is a minimal submanifold, its Gauss map is harmonic. The theorem then follows from the principal result of the paper, which is a Liouville theorem for certain harmonic maps between Riemannian manifolds, applied to the case where the target manifold is a Grassmannian.

Note that in the case of codimension 1, we have $m = 1$, hence $p = 1$ , $\kappa = 1$ , and (12) becomes

$$\beta_0 < \frac{1}{\cos(\pi/2)} = \infty\ ,$$

while

$$\det(g_{ij}) = 1 + |\nabla f|^2.$$

Thus, (13) is simply the condition that $\nabla f$ be bounded, and the result reduces to Moser's Theorem.

On the other hand, in the case $n = 2$, $m \geqslant 2$, corresponding to a two-dimensional surface of arbitrary codimension ( $\geqslant 2$ ), we have $p = 2$, $\kappa = 2$, and (12) becomes

$$\beta_0 < 2.$$

Denoting the solution (12) in vector form as

$$f(x_1,x_2) = (f_1(x_1,x_2),\ldots,f_m(x_1,x_2)) ,$$

we have

$$\det(g_{ij}) = 1 + \left|\frac{\partial f}{\partial x_1}\right|^2 + \left|\frac{\partial f}{\partial x_2}\right|^2 + \left|\frac{\partial f}{\partial x_1} \wedge \frac{\partial f}{\partial x_2}\right|^2.$$

Thus, (12), (13) imply

$$\sum_{i,j} \left( \frac{\partial f_i}{\partial x_j} \right)^2 < 3 ,$$

in other words, a very strong gradient bound. However, as indicated earlier, by using complex-variable techniques that are available when $n = 2$ instead of harmonic mappings, one can generally derive stronger results. Thus, in this instance one has the much more general theorem (see Osserman [38], p.290):

_If (2) is an entire solution of the minimal surface equation, with_

$n = 2$ *and if each* $f_k$ *is bounded above and below by linear functions of* $x_1, x_2$, *then each* $f_k$ *is itself linear*.

Another variant is the following theorem of Simon ([49], p. 46): *if* (2) *defines an entire solution of the minimal surface equation* (1), *where* $n = 2$, *and if all but one of the* $f_k$ *have bounded gradient, then all the* $f_k$ *are linear in* $x_1, x_2$.

One can in fact give a fairly concrete description of all entire solutions of the minimal surface equation for $n = 2$ in terms of entire holomorphic functions ([36], p.38). When also $m = 2$, this description is totally explicit ([36], p.42). One can, for example, easily deduce the following consequence:

*Let* (2) *be an entire solution of the minimal surface equation, where* $n = 2$ *and* $m = 2$. *If the corresponding surface has finite total curvature, then either* $f_1$ *and* $f_2$ *are linear, or else* $f_1 + if_2$ *is a complex polynomial in* $x_1 + ix_2$ *or in* $x_1 - ix_2$.

In a somewhat similar vein, there is the recent result of Micallef ([26], §5): *if the surface defined by an entire solution to the minimal surface equation for* $n = 2$ *and* $m = 2$ *is stable, then the surface is a holomorphic curve in* $\mathbb{C}^2$ *with respect to some orthogonal complex structure on* $\mathbb{R}^4$. This shows that the explicit examples of entire solutions ([36], p.42) referred to above define unstable surfaces in $\mathbb{R}^4$.

We end here our discussion of entire solutions of the minimal surface equation. In conclusion, let us note briefly some other important directions of recent work on the minimal surface equation.

**Regularity and singularities.**

*A solution of the codimension-one minimal surface equation in*

$D \setminus E$, <u>where</u> $D$ <u>is a bounded set, always extends to a solution in all of</u> $D$ <u>if</u> $E$ <u>is sufficiently small</u>. The strongest form of this result is due to Simon [48], who showed that $E$ can be an arbitrary set whose (n-1)-dimensional Hausdorff measure is zero. The remarkable part of this theorem is that no further assumptions, such as boundedness, need be made on the solution to conclude that it can be smoothly extended over E.

For higher codimension, no such result holds. There are even bounded solutions with isolated singularities [38]. However, when $n = 2$, an isolated singularity to which a function extends continuously is always removable [39], [15]. On the other hand, the equations (5) give an explicit example of a solution in all of $\mathbb{R}^4$ except the origin, extending continuously to the origin, but with a non-removable singularity there. For related results, see Harvey-Lawson [16]. (For some recent results and further references, see L. Simon, Annals of Math. 118 (1983), 525-571.)

**The Dirichlet problem.**

For codimension one the Dirichlet problem is quite thoroughly understood (see [11], [30], [40]). Uniqueness of solutions in a bounded domain $D$ follows from the maximum principle, while existence for arbitrary continuous boundary values holds if and only if the mean curvature of $\partial D$ is non-negative with respect to the inner normal. One can also consider generalized solutions of the Dirichlet problem when the boundary does not satisfy such a condition, and when the boundary function is not continuous. Again, very satisfactory results are available [12], [28], [23]. We note that a fundamental theorem for many of the results we have discussed is the <u>a priori</u> gradient bound of Bombieri, deGiorgi and Miranda [5]. (See also Trudinger [51].)

For regularity up to the boundary see Giusti [13] and

Lieberman [24], and for numerical methods, one has the papers of Hackbusch [14] and Concus, Golub and O'Leary [8] as well as further references cited there.

When we move from codimension one to higher codimension, the situation is entirely different. When the dimension n is 2, it is still true in arbitrary codimension that <u>a solution of the minimal surface equation exists in a bounded plane domain D taking on arbitrarily prescribed continuous boundary values if and only if D is convex</u> (see [36], Th. 7.2]. On the other hand, whenever the codimension is greater than one, <u>the solution may not be unique</u> ([21], Th. 5.1). Finally, in the general case (arbitrary dimension and codimension) <u>a solution of the Dirichlet problem may not exist even when D is the unit ball and the boundary values are the smoothest possible — quadratic polynomials</u> ([21], §6).

Clearly, the history of the minimal surface equation has been equally studded with surprises and successes. One can reasonably expect no less in the future.

**References**

[1] F. J. Almgren, Jr., Some interior regularity theorems for minimal surfaces and an extension of Bernstein's Theorem, Ann. of Math., 84 (1966), 277-292.

[2] S. Bernstein, Sur un théorème de géométrie et ses applications aux équations aux dérivées partielles du type elliptique, Comm. de la Soc. Math. de Kharkov (2) 15 (1915-1917), 38-45. (Also German translation: Über ein geometrisches Theorem und seine Anwendung auf die partiellen Differentialgleichungen vom elliptischen Typus, Math. Z. 26 (1927), 551-558.)

[3] L. Bers, Isolated singularities of minimal surfaces, Ann. of Math. (2) 53 (1951), 364-386.

[4] E. Bombieri, E. deGiorgi, and E. Giusti, Minimal cones and the Bernstein Theorem, Inventiones Math. 7 (1969), 243-269.

[5] E. Bombieri, E. deGiorgi, and M. Miranda, Una maggiorazione a priori relativa alle ipersuperfici minimali non parametriche, Arch. Rat'l. Mech. Anal. 32 (1969), 255-267.

[6] E. Bombieri and E. Giusti, Harnack's inequality for elliptic differential equations and minimal surfaces, Inventiones Math. 15 (1972), 24-46.

[7] S.-S. Chern, Minimal surfaces in an euclidean space of N dimensions, Differential and Combinatorial Topology, Princeton Univ. Press (1965), 187-198.

[8] P. Concus, G.H. Golub, and D.P. O'Leary, Numerical Solution of

Nonlinear Elliptic Partial Differential Equations by a Generalized Conjugate Gradient Method, Computing 19 (1978), 321-339.

[9] E. deGiorgi, Una extensione del teorema di Bernstein, Ann. Scuola Norm. Sup. Pisa 19 (1965), 79-85.

[10] W.H. Fleming, On the oriented Plateau problem, Rend. Circ. Mat. Palermo (2) 11 (1962), 69-90.

[11] D. Gilbarg and N.S. Trudinger, Elliptic Partial Differential Equations of Second Order, Grundlehren der mathematischen Wissenschaften 224, 2nd edition, Springer-Verlag, New York 1983.

[12] E. Giusti, Superfici cartesiane di area minima, Rend. Sem. Mat. Fis. Milano 40 (1970), 135-153.

[13] E. Giusti, Boundary behavior of non-parametric minimal surfaces, Indiana Univ. Math. J. 22 (1972), 435-444.

[14] W. Hackbusch, On the fast solutions of nonlinear elliptic equations, Numer. Math. 32 (1979), 83-95.

[15] R. Harvey and H. Blaine Lawson, Jr., Extending minimal varieties, Inventiones Math. 28 (1975), 209-226.

[16] R. Harvey and H.B. Lawson, Jr., A constellation of minimal varieties defined over the group $G_2$, Partial Differential Equations and Geometry. Proceedings of Park City Conference, edited by C.I. Byrnes, New York, M Dekker, 1979, pp. 167-187.

[17] E. Heinz, Über die Lösungen der Minimalflächengleichung, Nachr. Akad. Wiss. Göttingen, Math. Phys. Kl. II (1952), 51-56.

[18] S. Hildebrandt, J. Jost, and K.O. Widman, Harmonic mappings

and minimal submanifolds, Inventiones Math. 62 (1980), 269-298.

[19] E. Hopf, On S. Bernstein's theorem on surfaces z(x,y) of nonpositive curvature, Proc. Amer. Math. Soc. 1 (1950), 80-85.

[20] E. Hopf, On an inequality for minimal surfaces z = z(x,y) , J. Rat'l. Mech. Anal. 2 (1953), 519-522, 801-802.

[21] H.B. Lawson, Jr. and R. Osserman, Non-existence, non-uniqueness and irregularity of solutions to the minimal surface system, Acta. Math. 139 (1977), 1-17.

[22] H. Lebesgue, Intégrale, longueur, aire, Ann. Mat. Pura. Appl. (3) 7 (1902), 231-359.

[23] A. Lichnewsky, Solutions généralisées du problème des surfaces minimales pour des données au bord non bornées, J. Math. Pures Appl. (9) 57 (1978), 231-253.

[24] G.M. Lieberman, The quasilinear Dirichlet problem with decreased regularity at the boundary, Comm. Partial Differential Equations 6 (1981), 437-497.

[25] M. Meier, Removable singularities of bounded harmonic mappings and minimal submanifolds, preprint.

[26] M.J. Micallef, Stable minimal surfaces in euclidean space, MSRI preprint 011-83.

[27] E.J. Mickle, A remark on a theorem of Serge Bernstein, Proc. Amer. Math. Soc. 1 (1950), 86-89.

[28] M. Miranda, Dirichlet problem with $L^1$ data for the non-homogeneous minimal surface equation, Indiana Univ. Math. J. 24

(1974/75), 227-241.

[29] M. Miranda, Grafici minimi completi, Ann. Univ. Ferrara, Sez. VII - Sc. Mat. 22 (1977), 269-272.

[30] M. Miranda, Il problema di Dirichlet per l'equazione delle superfici minime, Rend. Sem. Mat. Fis. Milano 50 (1980), 117-121.

[31] J. Moser, On Harnack's Theorem for elliptic differential equations, Comm. Pure Appl. Math. 14 (1961), 577-591.

[32] J.C.C. Nitsche, Elementary proof of Bernstein's theorem on minimal surfaces, Ann. of Math. (2) 66 (1957), 543-544.

[33] J.C.C. Nitsche, Vorlesungen über Minimalflächen, Springer-Verlag, Berlin 1975.

[34] J.C.C. Nitsche, Minimal surfaces and partial differential equations, Studies in partial differential equations, 69-142, MAA Studies in Mathematics, 23, Mathematical Association of America, Washington, D.C., 1982.

[35] R. Osserman, On the Gauss curvature of minimal surfaces, Trans. Amer. Math. Soc. 96 (1960), 115-128.

[36] R. Osserman, A Survey of Minimal Surfaces, Van Nostrand Reinhold, New York, 1969.

[37] R. Osserman, Minimal varieties, Bull. Amer. Math. Soc. 75 (1969), 1092-1120.

[38] R. Osserman, Some properties of solutions to the minimal surface system for arbitrary codimension, Proc. Symp. Pure Math. 15, Global Analysis, Amer. Math. Soc., Providence 1970, pp. 283-291.

[39] R. Osserman, On Bers' Theorem on isolated singularities, Indiana Univ. Math. J. 23 (1973), 337-342.

[40] R. Osserman, Properties of solutions to the minimal surface equation in higher codimension, pp. 163-172 of Minimal Submanifolds and Geodesics, Proceedings of the Japan-United States Seminar on Minimal Submanifolds including Geodesics, Tokyo, 1977, Kaigai Publications, Tokyo 1978.

[41] J.T. Pitts, Existence and regularity of minimal surfaces on Riemannian manifolds, Princeton Math. Notes, 27, Princeton Univ. Press, Princeton 1982.

[42] T. Radó, Zu einem Satze von S. Bernstein über Minimalflächen im Grossen, Math. Z. 26 (1927), 559-565.

[43] E.A. Ruh, Asymptotic behavior of non-parametric minimal hypersurfaces, J. Diff. Geom. 4 (1970), 509-513.

[44] E.A. Ruh and J. Vilms, The tension field of the Gauss map, Trans. AMS 149 (1970), 569-573.

[45] R. Schoen and L. Simon, Regularity of stable minimal hypersurfaces in Riemannian manifolds, Comm. Pure Appl. Math. 34 (1981), 741-797.

[46] R. Schoen, L. Simon, and S.T. Yau, Curvature estimates for minimal hypersurfaces, Acta. Math. 134 (1975), 276-288.

[47] L. Simon, Remarks on curvature estimates for minimal hypersurfaces, Duke Math. J. 43 (1976), 545-553.

[48] L. Simon, On a theorem of deGiorgi and Stampacchia, Math. Z.

155 (1977), 199-204.

[49] L. Simon, A Hölder estimate for quasiconformal maps between surfaces in euclidean space, Acta Math. 139 (1977), 19-51.

[50] J. Simons, Minimal varieties in Riemannian manifolds, Ann. of Math. 88 (1968), 62-105.

[51] N.S. Trudinger, A new proof of the interior gradient bound for the minimal surface equation in n dimensions, Proc. Nat. Acad. Sci. U.S.A. 69 (1972), 821-823.

[52] S.-T. Yau, Survey on Partial Differential Equations in Differential Geometry, in Seminar on Differential Geometry, Annals of Math. Studies, No. 102, Princeton Univ. Press, Princeton, 1982, pp. 3-71.

# A Survey of Removable Singularities

By John C. Polking

Department of Mathematics
Rice University
P.O. Box 1892
Houston, TX 77251

Research partially supported by
NSF Grant MPS-75-05270

1. Introduction

Suppose $P(x,D) = \sum_{|\alpha| \leq m} a_\alpha(x) D^\alpha$ is a linear partial differential operator defined on an open set $\Omega \subset \mathbb{R}^n$, and that $A \subset \Omega$ is closed. Given a class $\mathcal{F}(\Omega)$ of distributions on $\Omega$, the set A is said to be removable for $\mathcal{F}(\Omega)$ and $P(x,D)$, if each $f \in \mathcal{F}(\Omega)$, which satisfies $P(x,D)f = 0$ in $\Omega \setminus A$ also satisfies $P(x,D)f = 0$ in $\Omega$.

The prototype of a removable singularity theorem is that of Riemann: if a function f is holomorphic in the punctured unit disk, and $f(z) = o(|z|^{-1})$ as z approaches zero, then f is holomorphic in the whole disk.

In general, a removable singularity theorem will assume a restriction on the growth of f near A, and a restriction on the size of A, and conclude that A is removable. The first general theorem of this type was proved by Bochner [B1]. Bochner's theorem is remarkable in that the restrictions on the growth of f and on the size of A depend only on the order of the operator P(x,D) and not on the type of the operator. Consequently, the theorem applies even to linear systems of differential operators such as d and $\bar{\partial}$. In fact, Bochner's method of proof can be extended even to some nonlinear equations.

It is our purpose here to give a survey of results on removable singularities. In §2 we first review the necessary facts about Hausdorff measure and capacity and then state the results for linear equations. In §3 we specialize to those results of interest in complex analysis, and state results about extension of holomorphic and plurisubharmonic functions, and about extension of positive d-closed (p,p) currents. Finally, in §4 we state a few results about removable singularities for some nonlinear equations. We do not try to be exhaustive here. Instead we specialize to some equations of geometric interest where the basic technique of Bochner yields some results of interest. We will not prove all of the results stated. Instead we will prove only a subset of the results, and give references to the literature for the rest.

## 2. Removable Singularities for Linear Equations

2.1 Let $C^k(\Omega)$ denote the space of functions on $\Omega$ whose derivatives up to order k are continuous. Let $C^{k,\delta}(\Omega)$, $o < \delta \leqslant 1$, denote the subset of $C^k(\Omega)$ consisting of those functions whose derivatives of order k satisfy a local Hölder condition of order $\delta$; i.e., for each compact set $K \subset \Omega$, and each multi-index $\alpha$ of length k, there is a constant C such that

$$|D^{\alpha}f(x) - D^{\alpha}f(y)| \leqslant C\,|x-y|^{\delta} \quad \text{if } x,y \in K.$$

Finally, $L^p_{k,loc}(\Omega)$ will denote the class of functions all whose derivatives up to order k belong to $L^p$ on compact subsets of $\Omega$.

2.2 To measure the size of A we introduce three set functions. For $o \leqslant d \leqslant n$, we define

$$\lambda_{d,\epsilon}(A) = \frac{\pi^{d/2}}{\Gamma(d/2+1)} \inf \Sigma\, r_i^d,$$

where the infinimum is over all coverings of A by balls $B_i$ of radius $r_i$. The d-dimensional <u>Hausdorff</u> <u>measure</u> of A is

$$\lambda_d(A) = \lim_{\epsilon\to 0^+} \lambda_{d,\epsilon}(A).$$

Hausdorff measure is a regular metric outer measure on $\mathbb{R}^n$, and consequently, $\lambda_d(A) = 0$ if and only if $\lambda_d(K) = 0$ for all compact set $K \subset A$. For $A \subset \Omega$, we will say that $\lambda_d(A)$ is <u>locally</u> <u>finite</u> <u>in</u> $\Omega$ if $\lambda_d(A \cap K) < \infty$ for all compact sets $K \subset \Omega$. For $d = n$, $\lambda_n$ coincides with the standard Legesque measure on $\mathbb{R}^n$.

Complete treatments of Hausdorff measure are contained in [C2] and [F]. We will need the following facts.

2.2.1 <u>Lemma</u>: If $\lambda_d(A) > 0$, then the space of measures $\mu$ with compact support contained in A, and which satisfy

$$|\mu|\,(B(a,r)) \leqslant Cr^d$$

for all balls B(a,r), and for some constant C, has infinite dimension.

This result is contained in the thesis of Frostman. A proof may be found in [C2] (Theorem II.1).

2.2.2 Lemma: Suppose $K \subset \mathbb{R}^n$ is compact. Given $d = n-mq$, and $\epsilon > 0$, there is a $\chi_\epsilon \in C_o^\infty(\mathbb{R}^n)$ with $\chi_\epsilon \equiv 1$ in a neighborhood of K and $\operatorname{supp} \chi_\epsilon \subset K_\epsilon = \{x \mid d(x,K) < \epsilon\}$, such that

$$\|D^\alpha \chi_\epsilon\|_q \leqslant C\, \epsilon^{m-|\alpha|} (\lambda_{n-mq}(K) + \epsilon)^{1/q}$$

for all $|\alpha| \leqslant m$, where C is a constant depending only on m and n.

For a proof see [HP1], §3.

If $A \subset \mathbb{R}^n$ is bounded, the d-dimensional lower Minkowski content of A is defined to be

$$M_d(A) = \liminf_{\epsilon \to 0^+} \epsilon^{d-n} \lambda_n(A_\epsilon),$$

where $A_\epsilon = \{x \mid d(x,A) < \epsilon\}$. The upper Minkowski content of A, $M^d(A)$, is defined similarly using lim sup. In general, we have $\lambda_d(A) \leqslant M_d(A) \leqslant M^d(A)$. There are examples ([K], [F]; also see 2.4.13) to show that the reverse inequalities are not true in general. However, $\lambda_d$, $M_d$, and $M^d$ all reduce to d-dimensional Lebesque measure on compact subsets of a smooth submanifold of $\mathbb{R}^n$ of dimension d. Notice that $M_d$ and $M^d$ are contents and not measures. In particular, these set functions are not countably additive. Minkowski content is discussed in [F].

For a compact set $K \subset \mathbb{R}^n$, $o \leqslant s < \infty$, and $1 < q < \infty$ we define the (s,q) capacity of K by

$$B_{s,q}(K) = \inf \{\|\varphi\|_{q,s} \mid \varphi \in C_o^\infty(\mathbb{R}^n) \text{ and } \varphi \geqslant 1 \text{ on } K\}.$$

The norm $\|\varphi\|_{q,s}$ is the norm in $L_s^q(\mathbb{R}^n)$ defined by $\|\varphi\|_{q,s} = \|g\|_q$, where $\varphi = G^s * g$, and $G^s$ is the Bessel potential of order s (i.e., $\hat{G}^s(\xi) = (1 + |\xi|^2)^{-s/2}$).

For $A \subset \mathbb{R}^n$ arbitrary, we define $B_{s,q}(A) = \sup B_{s,q}(K)$, the supremum being over compact subsets of A.

These capacities have been studied extensively (see [M], [R], [AM], [AP], [L1] and [L2]). We will mention only a few of their properties.

2.2.3 $B^{\cdot}_{s,q}$ is a capacity in the sense of Choquet; in particular, $B_{s,q}$ is countably subadditive:

$$B_{s,q}\left(\bigcup_{k=1}^{\infty} A_k\right) \leqslant \sum_{k=1}^{\infty} B_{s,q}(A_k)$$

2.2.4 If $1 < q < \infty$, $\frac{1}{p} + \frac{1}{q} = 1$, $o < s < \infty$, and $K \subset \mathbb{R}^n$ is compact, then

$$B_{s,q}(K) = \sup \mu(K)$$

where the supremum is over positive measures $\mu$ with supp $\mu \subset K$, and $\|\mu * G^s\|_p = \|\mu\|_{p,-s} \leqslant 1$.

2.2.5 For all compact sets K, $B_{s,q}(K) \sim C_{s,q}(K)$ where

$$C_{s,q}(K) = \inf \{\|\varphi\|_{q,s} \mid \varphi \in C_o^\infty(\mathbb{R}^n)\ \varphi \equiv 1 \text{ near } K\}.$$

Here $\sim$ means that the ratio is bounded above and below by nonzero, finite constants independent of K (see [L2], [AP]).

2.2.6 It follows easily from 2.2.5 that $B_{s,q}(A) = 0$ if and only if the only distribution in $L^p_{-s}(\mathbb{R}^n)$ with compact support contained in A is the zero distribution.

2.2.7 If $B_{s,q}(A) > 0$, then $\dim L^p_{-s}(\mathbb{R}^n)_A = \infty$ where $L^p_{-s}(\mathbb{R}^n)_A$ is the space of distributions in $L^p_{-s}(\mathbb{R}^n)$ with compact support contained in A (see [EP]).

2.2.8 Capacity and Hausdorff measure are closely related:

a) If $\lambda_{n-sq}(K) < \infty$, then $B_{s,q}(K) = 0$.

b) If $\lambda_{n-sq+\epsilon}(K) > 0$ for some $\epsilon > 0$, then $B_{s,q}(K) > 0$.

(see [M], [AM]; for s a positive integer, part a) follows easily from 2.2.2 and 2.2.5).

2.3 The problem of removable singularities for linear equations is put into proper perspective if we clearly separate it from the problem of regularity of solutions. That is, the solutions which we discuss will be weak solutions, or solutions in the sense of distributions. The regularity of such solutions is another problem which we will not discuss here.

Let

$$P(x,D) = \sum_{|\alpha| \leqslant m} a_\alpha(x) D^\alpha$$

be a linear partial differential operator with coefficients $a_\alpha$ which are matrices of functions in $C^\infty(\Omega)$. (This smoothness condition on the coefficients is needlessly restrictive for most of our results.) The formal adjoint of P(x,D) is defined by

$$^tP(x,D)\varphi = \sum_{|\alpha| \leqslant m} (-1)^{|\alpha|} D^\alpha({}^ta_\alpha \varphi).$$

We will let $d(z,A) = \inf \{ |x-y| \mid y \in A \}$. The pairing between $f \in \mathcal{D}'(\Omega)$ and $\varphi \in C_o^\infty(\Omega)$ will be denoted by $(f,\varphi) = f(\varphi)$.

To facilitate comparison of the results, we will put them in a table.

| | Condition on f | Condition on A | Remarks (see 2.4) |
|---|---|---|---|
| 2.3.1 | $f \in C^{0,\alpha}(\Omega)$, $0 < \alpha \leqslant 1$ | $\lambda_{n-m+\alpha}(A) = 0$ | 1, 2, 3, 4, 5 |
| 2.3.2 | $f \in C(\Omega)$ | $\lambda_{n-m}(A)$ is locally finite | 4, 5, 6 |
| 2.3.3 | $f \in L^{\infty}_{loc}(\Omega)$ | $\lambda_{n-m}(A) = 0$ | 4, 5, 7, 8 |
| 2.3.4 | $f \in L^{p}_{loc}(\Omega)$, $1 < p < \infty$, $1/p + 1/q = 1$ | $\lambda_{n-mq}(A)$ is locally finite | 4, 9 |
| | | $B_{m,q}(A) = 0$ | 2, 5, 10, 11 |
| 2.3.5 | $f \in L^{1}_{loc}(\Omega)$ and $f(x) = O(d(x,A)^{-a})$ uniformly on compact subsets of A | $M_{n-m-a}(K) = 0$ for all compact sets $K \subset A$ | 12, 13 |
| | $f \in L^{1}_{loc}(\Omega)$ and $f(x) = o(d(x,A)^{-a})$ uniformly on compact subsets of A. | $M_{n-m-a}(K) < \infty$ for all compact sets $K \subset A$ | |

Sufficient Conditions for Removability

$$P(x,D)f = 0 \text{ in } \Omega\setminus A$$

$$P(x,D) = \sum_{|\alpha|\leqslant m} a_\alpha(x)D^\alpha, \quad a_\alpha \in C^\infty(\Omega)$$

We will give a proof for 2.3.2; see the remarks in 2.4 for references to the literature for the other results. To show that $P(x,D)f = 0$ in $\Omega$, it suffices to show that $(P(x,D)f,\varphi) = 0$ for all $\varphi \in C_o^\infty(\Omega)$. Let $K = A \cap \operatorname{supp} \varphi$, and for $\epsilon > 0$ choose $\chi_\epsilon$ by Lemma 2.2.2. Then

$$(P(x,d)f,\varphi) = (P(x,D)f,\chi_\epsilon\varphi) = (f,{}^tP(x,D)(\chi_\epsilon\varphi))$$

$$= \sum_{|\alpha|\leqslant m} (fD^\alpha\chi_\epsilon,\varphi_\alpha)$$

where $\varphi_\alpha \in C_o^\infty(\Omega)$. By Lemma 2.2.2 with $q = 1$,

$$|(f\varphi_\alpha,D^\alpha\chi_\epsilon)| \leqslant C\,\|f\|_\infty\,\epsilon^{m-\alpha}\,(\lambda_{n-m}(K) + \epsilon)$$

Hence, $(f\varphi_\alpha,D^\alpha\chi_\epsilon) \to 0$ if $|\alpha| < m$. For $|\alpha| = m$, we notice that $D^\alpha\chi_\epsilon$ is a bounded net in $L^1(\Omega)$, and hence has a subsequence which is weakly convergent in $C(\Omega)'$. Since $D^\alpha\chi_\epsilon$ converges to zero in $\mathfrak{D}'(\Omega)$, the subsequence converges weakly to zero in $C(\Omega)'$. Multiplication by the continuous function f is continuous in $C(\Omega)'$. Hence, a subsequence of the net $fD^\alpha\varphi_\epsilon$ converges weakly to zero in $C(\Omega)'$. It follows that $(P(x,D)f,\varphi) = 0$.

This is a typical proof of a removable singularity result. The key point is the proof of the existence of the cut-off function $\chi_\epsilon$ with the right properties. This is provided by Lemma 2.2.2 for 2.3.2, 2.3.3, and for the Hausdorff measure statements in 2.3.4. It is provided by 2.2.5 for 2.3.4. For 2.3.1 see [HP1].

2.4 Remarks

2.4.1 For the standard Euclidean Laplacian

$\Delta = \sum_{j=1}^{n} \left[\frac{\partial}{\partial x^j}\right]^2$, this result was proved by Carleson ([C1], [C2]). Carleson also showed that the condition on A is necessary for removability in this case.

2.4.2 If $P(x,D)$ is a determined elliptic operator in $\Omega$, then this condition on A is necessary and sufficient for removability relative to $\Omega'$ for any open set $\Omega' \subset\subset \Omega$. The necessity may be seen as follows. The determined elliptic operator P(x,D) has a parametrix Q(x,D); i.e., Q(x,D) is a pseudodifferential operator of order $-m$ on $\Omega$ such that

$$P \circ Q = I + S_1$$

$$Q \circ P = I + S_2$$

where $S_1$ and $S_2$ are operators that map $\mathcal{E}'(\Omega)$ into $C^\infty(\Omega)$. It follows that if $1 < q < \infty$ and $\frac{1}{p} + \frac{1}{q} = 1$, there is a constant C such that

$$\|\varphi\|_{q,m} \leqslant C\ [\|{}^tP(x,D)\varphi\|_q + \|\varphi\|_q],\ \varphi \in C_o^\infty(\Omega').$$

Then, if $L^q_{m,o}(\Omega')$ denotes the closure of $C_o^\infty(\Omega')$ in $L^q_m(\Omega')$, we see that ${}^tP(x,D)$: $L^q_{m,o}(\Omega') \to L^q(\Omega')$ has closed range and a finite dimensional null space. Consequently, P(x,D): $L^p(\Omega') \to L^p_{-m}(\Omega')$ has closed range of finite codimension ($L^p_{-m}(\Omega')$ is defined to be the dual of $L^q_{m,o}(\Omega')$). If $A \subset \Omega'$ satisfies $B_{m,q}(A) > 0$, then by 2.2.7, $L^p_{-m}(\mathbb{R}^n)_A \subset L^p_{-m}(\Omega')$ has infinite dimension. Hence, there is an $f \in L^p(\Omega')$, such that $0 \neq P(x,D)f \in L^p_{-m}(\mathbb{R}^n)_A$, so A is not removable for $L^p_{loc}(\Omega')$ and P(x,D), and the condition in 2.3.4 is necessary for P(x,D) relative to $\Omega'$.

If $\mu$ is a measure with compact support such that

$$* \qquad |\mu|\ (B(a,r)) \leqslant Cr^{n-m+\alpha} \text{ for all balls}$$

then it is easily shown that $G^m * \mu \in C^{o,\alpha}(\mathbb{R}^n)$. Hence, in particular $\mu \in L^p_{-m}(\mathbb{R}^n)$ for any p, $1 < p < \infty$.

If $\lambda_{n-m+\alpha}(A) > 0$, then $B_{m,q}(A) > 0$, so we may apply the argument of the previous paragraph together with 2.2.1 to conclude that there is an $f \in L^p(\Omega')$ such that $0 \neq P(x,D)f = \mu$, where $\mu$ is a measure with compact support in A which satisfies *. Since P(x,D) is elliptic, $f \in C^\infty(\Omega' \setminus \text{supp } \mu)$. Hence, if $\chi \in C_o^\infty(\Omega')$ is identically equal to 1 near supp $\mu$, $(1-\chi)f \in C^\infty(\Omega')$,

$$P(x,D)\chi f = P(x,D)f - P(x,D)[(1-\chi)f)]$$

$$= \mu - P(x,D)[(1-\chi)f].$$

Applying Q(x,D), we have

$$\chi f + S_2\chi f = Q(x,D)\mu - Q(x,D)P(x,D)\ [(1-\chi)f].$$

Consequently, $f - Q(x,D)\mu \in C^\infty(\Omega')$. Since Q(x,D) is a pseudodifferential operator of order $-m$, $Q(x,D)\ (1-\Delta)^{m/2}$ is a pseudodifferential operator of order 0. Since $\mu$ satisfies *, $(1-\Delta)^{-m/2}\mu = G^m * \mu \in C^{o,\alpha}(\mathbb{R}^n)$. Hence, by Theorem XI.2.1 in [T], $Q(x,D)\mu \in C^{o,\alpha}(\Omega)$ and, therefore, $f \in C^{o,\alpha}(\Omega')$. The argument can be made simpler by noticing (see Seeley [SE]) that the operator Q(x,D) has a kernel Q(x,y) which satisfies the estimates

$$|Q(x,y)| \leq \text{Const.}\ |x-y|^{m-n}$$

$$|D^\alpha Q(x,y)| \leq \text{Const.}\ |x-y|^{m-n-|\alpha|}$$

at least for x,y restricted to compact subsets of $\Omega$. It follows from the second inequality that

$$|Q(x,z) - Q(y,z)| \leq \text{Const.}\ |x-z|^{m-n-1} \cdot |x-y|$$

if $|x-y| \leq \frac{1}{2}|x-z|$. It is now a standard argument to show that if $\mu$ satisfies *, then

$$Q(x,D)\mu = \int Q(x,z)\ d\mu(z)$$

belongs to $C^{o,\alpha}$ (see the proof of Theorem VII.2 in [C2] for example).

2.4.3 If $f \in C^{k,\alpha}(\Omega)$, $0 \leq k < m$, $0 < \alpha \leq 1$, then the condition on A becomes $\lambda_{n-m+k+\alpha}(A) = 0$.

2.4.4 This result is proved in [HP1].

2.4.5 Necessary and sufficient conditions for a wide class of operators are given in [HP2].

2.4.6 For the Laplacian $\Delta$ in $\mathbb{R}^n$, and for the Cauchy-Riemann operator $\frac{\partial}{\partial \bar{z}}$ in $\mathbb{C}$, necessary and sufficient conditions are classical. For $\Delta$, it is necessary and sufficient that A be polar, or equivalently, that the Wiener capacity of A be zero. For $\frac{\partial}{\partial \bar{z}}$, it is necessary and sufficient that the continuous analytic capacity (AC capacity) of A vanish.

2.4.7 If $\lambda_{n-m}(A)$ is locally finite, it follows that P(x,D)f is a measure supported in A (see [HP1]).

2.4.8 For $\Delta$, it is necessary and sufficient that A be polar. For $\frac{\partial}{\partial \bar{z}}$, it is necessary and sufficient that A have zero analytic capacity.

2.4.9 This result was first proved by Carleson for $\Delta$ and for compact sets A ([C2]). For elliptic operators of second order this result is due to Serrin [S1].

2.4.10 This result was proved by Littman ([L1] and [L2]). The connection between capacity and Hausdorff measure was not made until later ([M], also see Lemma 3.2 in [HP1]).

2.4.11 If $f \in L^p_{s,loc}(\Omega)$, then $B_{m-s,q}(A) = 0$ is sufficient. This condition is also necessary for determined elliptic operators on sets $\Omega' \subset\subset \Omega$.

2.4.12 Bochner proved the second result under the stronger hypothesis that $M^{n-m-a}(K) < \infty$ for all compact sets $K \subset A$. (See [B1] and also [HP1]). The proofs of the stronger results stated here are very similar to Bochner's proof. It suffices to show that $(P(x,D)f,\varphi) = 0$ for all $\varphi \in C_o^\infty(\Omega)$. Let $K = A \cap \operatorname{supp} \varphi$. Choose $\psi \in C_o^\infty(\mathbb{R}^n)$ with $\operatorname{supp} \psi \subset B(0,1)$, and $\int \psi(x)dx = 1$. For $\epsilon > 0$, define

$$\chi_\epsilon(x) = \epsilon^{-n} \int_{K_{2\epsilon}} \psi\left(\frac{x-y}{\epsilon}\right) dy$$

Then $\chi_\epsilon(x) \equiv 1$ in $K_\epsilon$, $\operatorname{supp} \chi_\epsilon \subset K_{3\epsilon}$, and $|D^\alpha \chi_\epsilon(x)| \leq C_\alpha \epsilon^{-|\alpha|}$. Since $\chi_\epsilon \equiv 1$ near K,

$$(P(x,D)f,\varphi) = (P(x,D)f,\chi_\epsilon\varphi) = (f,{}^tP(x,D)(\chi_\epsilon\varphi))$$

$$= \sum_{|\alpha|\leqslant m} (f,\varphi_\alpha D^\alpha\chi_\epsilon),$$

where $\varphi_\alpha \in C_o^\infty(\Omega)$. For $\alpha = 0$,

$$|(f,\varphi_0\chi_\epsilon)| \leqslant C\int_{K_{3\epsilon}} |f(x)|\,dx \to 0 \text{ as } \epsilon\to 0,$$

since $f \in L^1_{loc}(\Omega)$. For $\alpha \neq 0$, notice that $D^\alpha\chi_\epsilon \equiv 0$ in $K_\epsilon$. Let

$$\sup_{x\in K_{3\epsilon}\setminus K_\epsilon} |\varphi_\alpha(x)f(x)| = c_\epsilon\epsilon^{-a}$$

Then

$$|(f,\varphi_\alpha D^\alpha\chi_\epsilon)| \leqslant C_\alpha c_\epsilon\epsilon^{-a-|\alpha|}\,\lambda_n(K_{3\epsilon})$$

$$= C_\alpha c_\epsilon\epsilon^{m-|\alpha|}\cdot\epsilon^{-a-m}\,\lambda_n(K_{3\epsilon})$$

Under either of the two sets of hypotheses, there is a sequence $\epsilon_j\to 0$, such that $c_{\epsilon_j}\cdot\epsilon_j^{-a-m}\,\lambda_n(K_{3\epsilon_j})\to 0$. Consequently, $(P(x,D)f,\varphi) = 0$, and A is removable.

2.4.13 The conditions involving Minkowski content are certainly weak. Consider the following example. For $k = 0, 1, 2, \ldots$, let $A_k = \{p_k^j\}$ be a finite set such that $A_k \subset \{x \in \mathbb{R}^n \mid |x| \leqslant 2^{-k}\}$ and

$$(A_k)_\epsilon = \bigcup_j B(p_k^j,\epsilon) \supset \{x \in \mathbb{R}^n \mid |x| \leqslant 2^{-k}\}$$

if $\epsilon \geqslant 2^{-k^2}$. Then if $A = \{0\} \cup \left[\bigcup_{k=0}^{\infty} A_k\right]$, A is a countable compact set with the origin its only limit point. If $2^{-(k+1)^2} \leqslant \epsilon \leqslant 2^{-k^2}$ then

$$\epsilon^{d-n} \lambda_n (A_\epsilon) \geq 2^{(n-d)k^2-nk}$$

Consequently, $M_d(A) = \infty$ for all $d < n$.

On the other hand, suppose $P(x,D)$ is a differential operator of order $m < n$ defined in $\Omega$, a neighborhood of the origin. If $f \in L^1_{loc}(\Omega)$, $f(x) = o(d(x,A)^{m-n})$ uniformly on compact subsets of $\Omega$, and if $P(x,D)f = 0$ in $\Omega \setminus A$, then $P(x,D)f = 0$ in $\Omega$. This can be seen by applying 2.3.5 first in a neighborhood of each isolated point of A, and finally to 0, the limit point of A.

It might be possible to replace Minkowski content in 2.3.5 by Hausdorff measure, but even this condition seems to be less than optimal, since in the previous example $\lambda_0(A) = \infty$. On the other hand, if a is an integer, $0 < a \leq n-m$, and A is restricted to be a subset of a submanifold of dimension $n-m-a$, then the conditions on f and on A in 2.3.5 are optimal for elliptic operators as is easily seen. On the other hand if $a = 0$, stronger results are sometimes possible; see [HP1], Section 6.

## 3. Removable Singularities in Complex Analysis

In complex analysis problems of removable singularities arise for holomorphic functions, extension of plurisubharmonic functions, and the extension of positive, d-closed (p,p) currents. A survey of results in this direction through 1975 is contained in [HP3].

3.1 The problem of removable singularities for holomorphic functions is the same as for any other linear operator as discussed in the previous section. Indeed the results of §2.3 with $m = 1$ all apply and are of interest. However, in several complex variables, there are a class of new results that are due to the fact that the Cauchy-Riemann equations are an overdetermined elliptic system.

| | Condition on f | Condition on A | Remarks (see 3.2) |
|---|---|---|---|
| 3.1.1 | None | A compact<br>$\Omega \setminus A$ connected | 1 |
| 3.1.2 | None | $\lambda_{2n-2}(A) = 0$ | 2 |
| 3.1.3 | None | A is a real submanifold of $\Omega$ of dimension 2n-2 and $\dim_{\mathbb{C}} H_p(A) = n-2$ for all $p \in A$ | 3 |
| 3.1.4 | $f \in \mathcal{O}(\omega)$<br>for some open set $\omega$ with $\omega \cap A \neq 0$ | A is a connected complex variety in $\Omega$ | 4 |

Sufficient Conditions for Removability

$f \in \mathcal{O}(\Omega \setminus A)$, $\Omega \subset \mathbb{C}^n$, $n \geqslant 2$

3.2 Remarks

1. This is the Hartog's Theorem. See [H], Theorem 2.3.2 for a proof. This phenomenon is characteristic of elliptic overdetermined systems; see [B2] and [EH].

2. This result is due to Shiffman [SH]. See [HP3] for a proof.

3. This result is due to Cegrell [CE].

4. This result is classical; see [HP3] for a proof.

3.3 The extension of plurisubharmonic functions is slightly different than the removable singularity problems we have discussed until now. What is involved here is the extension of a partial differential inequality instead of a homogeneous equation.

A function (distribution) u defined on an open set $\Omega \subset \mathbb{C}^n$ is said to be plurisubharmonic if

$$i\partial\bar{\partial}u = i \sum \frac{\partial^2 u}{\partial z^i \partial \bar{z}^j} dz^i \wedge d\bar{z}^j$$

is a positive (1.1) current. A (p,p) current $\omega$ is positive if $\omega \wedge i\alpha_1 \wedge \bar{\alpha}_1 \wedge i\alpha_2 \wedge \bar{\alpha}_2 \wedge \ldots \wedge i\alpha_{n-p} \wedge \bar{\alpha}_{n-p}$ is a positive measure for all choices of smooth (1,0) forms $\alpha_1, \ldots, \alpha_{n-p}$. A good discussion of plurisubharmonic functions and positive currents is contained in Lelong [LE] and in Vladimirov [V].

The basic assumption that we make is that u is plurisubharmonic in $\Omega \setminus A$, where A is a closed subset of $\Omega$. The conclusion under the indicated hypotheses is that u has an extension $\tilde{u}$ which is plurisubharmonic in $\Omega$.

| | Condition on u | Condition on A | Remarks (see 3.4) |
|---|---|---|---|
| 3.3.1 | None | $\lambda_{2n-2}(A) = 0$ | 1 |
| 3.3.2 | None | A is a real submanifold of $\Omega$ of dimension 2n-2, and $\dim_{\mathbb{C}} H_p(A) = n-2$ for all $p \in A$ | 2 |
| 3.3.3 | u is plurisubharmonic in $\omega$ for some open set $\omega$ with $\omega \cap A \neq 0$ | $A \subset \mathbb{C}^n$ is proper complex subvariety | 3 |
| 3.3.4 | $u \in L^{\infty}_{loc}(\Omega)$ | A is polar | 4 |
| 3.3.5 | $u \in C^{o,\alpha}(\Omega)$ | $\lambda_{2n-2+\alpha}(A) = 0$ | 5 |

Sufficient Conditions for Extension

$i\partial\bar{\partial} u \geq 0$ in $\Omega \setminus A$

3.4 Remarks

1. Due to Shiffman [S]. A proof is contained in [HP3].

2. Due to Cegrell [CE]. $H_p(A)$ is the holomorphic tangent space of the submanifold A. It is defined to be $H_p(A) = T_p(A) \cap (iT_p(A))$ where $T_p(A)$ is the ordinary (real) tangent space.

3. Due to Siu [SI]. Siu's proof uses the Hörmander $L^2$ estimates. A more elementary proof is contained in [HP3].

4. In the special case when A is a complex subvariety, this result is due to Grauert and Remmert [GR]. The result as stated is due to Lelong [LE]. A set A is <u>polar</u> if there is a subharmonic function $\varphi$ on $\mathbb{C}^n = \mathbb{R}^{2n}$ for which $\varphi(z) = -\infty$ for all $z \in A$. Compare this result with 2.3.3 and 2.4.8

5. A proof is in [HP3]. Compare this result with 2.3.1.

3.5 If positive (p,p) current u is expressed as a form with distribution coefficients, $u = \sum u_{IJ} dz^I \wedge d\bar{z}^J$, then the coefficients $u_{IJ}$ are in fact measures (see [LE]). If u is defined in $\Omega \setminus A$, and each coefficient assigns a finite value to $K \cap (\Omega \setminus A)$ for every compact set $K \subset \Omega$, then we can define an extension $\tilde{u}_{IJ}$ of each coefficient to $\Omega$ by setting $\tilde{u}_{IJ}(K) = u_{IJ}\ (K \cap (\Omega \setminus A))$ for each $K \subset \Omega$. The extension $\tilde{u} = \sum \tilde{u}_{IJ} dz^I\ d\bar{z}^J$ obtained in this way is called the <u>trivial extension of u across A by zero</u>.

An extension theorem for positive, d-closed, (p,p) currents has the following form: u is assumed to be positive, and d-closed in $\Omega \setminus A$; then under suitable restrictions on u near A and on A, we conclude that the trivial extension of u by zero across A exists and furthermore is positive and d-closed in $\Omega$. The positivity is, of course, trivial; but the existence and the d-closed condition are not.

These results are motivated by the most interesting special case. Suppose $V \subset \Omega \setminus A$ is a pure k-dimensional subvariety. Then with $p = n - k$, [V] will denote the (p,p) current of integration forms over the regular points of V; i.e., $[V](\varphi) = \int_{\text{Reg } V} \varphi$. [V] is positive and d-closed. An extension theorem guarantees that $\tilde{u}$, the trivial extension of [V] by zero across A exists and is d-closed and positive in $\Omega$, and furthermore it is easily seen that $\tilde{u}$ is locally rectifiable. From the structure theorem of James King [KI] we can now conclude that $\bar{V} = \text{supp}\ \tilde{V}$ is a pure k-dimensional subvariety of $\Omega$. Such results on extension of varieties across exceptional sets preceded the results for positive, d-closed currents.

| | Conditions on u | Conditions on A | Remarks (see 3.6) |
|---|---|---|---|
| 3.5.1 | None | $\lambda_{2k-1}(A) = 0$ | 1 |
| 3.5.2 | u has an extension as a positive d-closed (p,p) current to some open set $\omega$ with $\omega \cap A \neq 0$ | A is a proper irreducible subvariety of $\Omega$ of dimension k | 2 |
| 3.5.3 | u has locally finite mass across A | A is a closed complete pluripolar set | 3 |
| 3.5.4 | None | A is a submanifold and $\dim_{\mathbb{C}} H_z(A) < k-1$ for all $z \in A$ | 4 |

Sufficient Conditions for $\tilde{u}$ to Exist and be d-Closed
u Positive, d-Closed, (p,p) Current on $\Omega \setminus A$, $p+k = n$

3.6 Remarks

1. The special case when $u = [V]$ and the exceptional set A is a subvariety is due to Remmert and Stein [RS]. When $u = [V]$, and $\lambda_{2k-1}(A) = 0$, the result was proved by Shiffman [SH]. The general result as stated is due to Harvey [H1].

2. Again the special case when $u = [V]$ is due to Remmert and Stein and is usually called the Remmert Stein Theorem ([RS]). The result as stated is due to Siu [SI]. A more elementary proof is contained in [HP3].

3. A subset $A \subset \Omega$ is a <u>complete</u> <u>pluripolar</u> <u>set</u> if there is a plurisubharmonic function $\varphi$ on $\Omega$ such that $A = \{z \in \Omega \mid \varphi(z) = -\infty\}$. There were several precursers of this theorem all of which assumed that A was a subvariety. When $u = [V]$ and $\dim A = \dim V$, the result is due to Stoll [ST]. For u a positive d-closed (p,p) current and $\dim A = k = n - p$, the result is in [HP3]. Bishop proved the result when A was assumed to be a hypersurface and $u = [V]$ where the dimension of V was unrestricted [BI]. The extension of Bishop's result to general d-closed, (p,p) currents was made by Skoda [SK]. When A is a hypersurface and u is a positive, d-closed, (1.1) current, a stronger result is contained in Theorem 2.1 in [HP3].

The result as stated is due to El Mir [EM]. We will give a proof of this result which is due to Siboney [SY].

Suppose first that u is a positive, d-closed, (p,p) current in $\Omega \setminus A$. Let $|u|$ denote the total variation measure of a current with measure coefficients. The assumption that we are making on u is that $|u|(K \cap (\Omega \setminus A)) < \infty$ for all compact sets $K \subset \Omega$. The existence of $\tilde{u}$, the extension of u across A by zero is, of course, immediate.

<u>Lemma 1</u>: (See [CLN].) For every compact set $K \subset \Omega$ there is a constant $C_1(K)$ such that

$$|u \wedge \partial\bar{\partial}\psi|(K) \leq C_1(K) \|\psi\|_\infty$$

for infinitely differentiable, plurisubharmonic functions $\psi$ defined in $\Omega$ which vanish in neighborhood of A.

Proof: Suppose not. Then there is a sequence $\psi_\nu$ of such functions such that $\|\psi_\nu\|_\infty \to 0$ but $|u\wedge\partial\bar\partial\psi_\nu|(K) \geq 1$ for all $\nu$. But then $\psi_\nu u \to 0$ in the distribution sense in $\Omega$ and it follows that $i\partial\bar\partial(\psi_\nu u) = i\partial\bar\partial\psi_\nu\wedge u \to 0$ weakly. However, $i\partial\bar\partial\psi_\nu\wedge u$ is a sequence of positive (p+1,p+1) currents, and for such objects weak convergence to zero implies strong convergence to zero, contradicting the hypothesis that $|\partial\bar\partial\psi_\nu\wedge u|(K) \geq 1$.

Lemma 2: For every compact set $K \subset \Omega$, there is a constant $C_2(K)$ such that

$$|i\partial\psi\wedge\bar\partial\psi\wedge u|\,(K \cap \{z \mid 1/4 \leq \psi(z) \leq 1/2\}) \leq C_2(K)$$

for all infinitely differentiable plurisubharmonic functions $\psi$ in $\Omega$ which vanish in a neighborhood of A and which satisfy $\|\psi\|_\infty \leq 1$.

Proof: Choose a function $\chi \in C^\infty(\mathbb{R})$ such that

$$0 \leq \chi(t) \leq t,\ 0 \leq \chi'(t) \leq 1,$$

$$\chi''(t) \geq 0 \text{ for all } t \in \mathbb{R},\ \chi''(t) \equiv c \text{ for } 1/4 \leq t \leq 1/2$$

Then

$$c\ |i\partial\psi\wedge\bar\partial\psi\wedge u|(K\cap\{z\mid 1/4\leq\psi(z)\leq 1/2\}) \leq |\chi''(\psi)i\partial\psi\wedge\bar\partial\psi\wedge u|(K)$$

Since

$$i\partial\bar\partial\chi(\psi)\wedge u = i\chi''(\psi)\partial\psi\wedge\bar\partial\psi\wedge u + i\chi'(\psi)\partial\bar\partial\psi\wedge u$$

$$|\chi''(\psi)i\partial\psi\wedge\bar\partial\psi\wedge u|(K) \leq |i\partial\bar\partial\chi(\psi)\wedge u|(K) + |\chi'(\psi)\partial\bar\partial\psi\wedge u|\ (K)$$

Applying Lemma 1 to u and to $\chi'(\psi)u$ we have

$$|i\partial\psi\wedge\bar\partial\psi\wedge u|\ (K \cap \{z \mid 1/4 \leq \psi(z) \leq 1/2\}) \leq 2c^{-1}C_1(K)$$

Proof of 3.5.3: Let $\varphi$ be a plurisubharmonic function in $\Omega$ such that $A = \{z \in \Omega \mid \varphi(z) = -\infty\}$. Then $\varphi$ is bounded above on any open set $\Omega' \subset\subset \Omega$. With such an $\Omega'$ fixed we may assume that $\varphi(z) \leq 0$ on $\Omega'$. Then $\exp(\varphi(z)/\nu)$ is plurisubharmonic in $\Omega$, vanishes on A, and satisfies $0 \leq \exp(\varphi(z)/\nu) \leq 1$ on $\Omega'$. Let $\theta$ be a non-decreasing convex function such that $\theta(t) = 0$ if $t \leq 1/2$, and $\theta(t) \geq 1$ if $t \geq 1$. Let $\tilde{\psi}_\nu(z) = \theta(\exp(\varphi(z)/\nu))$. Then $\tilde{\psi}_\nu$ is plurisubharmonic in $\Omega'$, vanishes near A, satisfies $0 \leq \tilde{\psi}_\nu(z) \leq 1$ in $\Omega'$, and $\tilde{\psi}_\nu(z) \to 1$ if $z \in \Omega' \setminus A$. Let $\psi_\nu$ denote a suitable smoothing of $\tilde{\psi}_\nu$. Then $\psi_\nu$ has all of the properties of $\tilde{\psi}_\nu$ and, in addition, $\psi_\nu$ is infinitely differentiable in $\Omega'$.

Now choose $\theta \in C^\infty(\mathbb{R})$ satisfying $\theta(t) = 0$ $t \leq -1/4$, $\theta(t) \equiv 1$ $t \geq 1/2$, $\theta'(t) \geq 0$ and set $\varphi_\nu = \theta(\psi_\nu)$. Clearly, $\tilde{u} = \lim_{\nu\to\infty} \varphi_\nu u$. Hence, $\bar{\partial}\tilde{u} = \lim_{\nu\to\infty} \bar{\partial}\varphi_\nu \wedge u$. It suffices to show that

$\bar{\partial}\tilde{u} = 0$, since $\partial\tilde{u} = \overline{\bar{\partial}\tilde{u}}$. To this end it suffices to show that

$$\langle \bar{\partial}u, \omega \wedge \eta \rangle = 0$$

for every smooth compactly supported (1,0) form $\eta$ and every smooth positive (n-p-1,n-p-1) form $\omega$. Since $\bar{\partial}\tilde{u} = \lim_{\nu\to\infty} \bar{\partial}\varphi_\nu \wedge u$, we examine

$$\langle \bar{\partial}\varphi_\nu \wedge u, \omega \wedge \eta \rangle = \langle \theta'(\psi_\nu)\bar{\partial}\psi_\nu \wedge u, \omega \wedge \eta \rangle$$

$$= \langle \theta'(\psi_\nu)\omega \wedge u, \bar{\partial}\psi_\nu \wedge \eta \rangle$$

Since $\theta'(\psi_\nu)\omega \wedge u$ is a positive (n-1,n-1) current, it follows from the Cauchy-Schwartz inequality that

$$|\langle \theta'(\psi_\nu)\omega \wedge u, \bar{\partial}\psi_\nu \wedge \eta \rangle|^2 \leq \langle \theta'(\psi_\nu)\omega \wedge u, i\partial\psi_\nu \wedge \bar{\partial}\psi_\nu \rangle$$

$$\cdot \langle \theta'(\psi_\nu)\omega \wedge u, i\eta \wedge \bar{\eta} \rangle$$

From Lemma 2 we conclude that

$$|\langle \bar{\partial}\varphi_\nu \wedge u,\ \omega\wedge\eta\rangle|^2 \leqslant C_2(K)\ \langle \theta'(\psi_\nu)\omega\wedge u,\ i\eta\wedge\bar{\eta}\rangle$$

$$\longrightarrow 0$$

Hence, $\bar{\partial}\tilde{u} = 0$.

4. This, again, is a result of El Mir. We refer the reader to his thesis [EM] for a proof. Earlier results in this direction are due to Becker [BE], Funahashi [FU] and Cirka [CI].

3.7 Finally, we mention the following result of Harvey [H2] on removable singularities in cohomology classes.

Theorem: Suppose A is a closed subset of $\Omega$ open in $\mathbb{C}^n$. If $\lambda_{2(n-k)}(A) = 0$, then $H^p(\Omega\setminus A, \mathcal{F}) \cong H^p(\Omega,\mathcal{F})$ for $p < k$ and for any locally free analytic sheaf $\mathcal{F}$.

See [H2] for a proof and applications.

4. Selected Results for Non-Linear Equations

We will be interested primarily in results for some specific non-linear equations. Our treatment can in no way be considered exhaustive. On the other hand, it should be pointed out that results of a general nature for quasilinear elliptic operators are contained in [S2], and in [MA]; for quasilinear elliptic systems see [ME1].

4.1 Harmonic maps. Before stating the results on removable singularities for harmonic maps, notice that the notions $\lambda_d(A) = 0$, $\lambda_d(A)$ is locally finite, $B_{s,q}(A) = 0$, are all differential invariants; and, consequently, make sense on smooth manifolds.

Let M be a Riemannian manifold of dimension m and metric g. Let N be another, of dimension n and metric h.

A map $f \in L^2_{1,loc}(M, N)$ is weakly harmonic if

$$(1) \qquad 0 = \int_M \langle df,\ \nabla\varphi\rangle\ v_g$$

for all compact variations $\varphi$, where $\nabla$ denotes the appropriate covariant differential and $v_g$ the volume element of the Riemannian metric g on M. We say that f is harmonic if it is weakly harmonic and smooth. To simplify coordinate calculations we express (1) as follows (see [ES, Chapter 2] and [HA]): Embed N in a finite dimensional vector space V. Take an open tubular neighborhood U of N in V; we can assume that U is realized as a Riemannian disc bundle, and that V is endowed with a complete Riemannian metric extending that of U. In a chart of M (1) takes the form

$$(2)\qquad 0=-\int g^{ij}f_i^{\gamma}\varphi_j\sqrt{\det g}\,dx+\int g^{ij}\pi^{\gamma}_{\alpha\beta}f_i^{\alpha}f_j^{\beta}\varphi\sqrt{\det g}\,dx\,(1\leqslant\gamma\leqslant n).$$

Here $(g^{ij})$ is the inverse of the matrix $(g_{ij})$, $f_i^{\alpha}=\dfrac{\partial f^{\alpha}}{\partial x^{i}}$, $\varphi_j$ is the covariant derivative of $\varphi$ in the $x^j$ direction, and $(\pi^{\alpha}_{\alpha\beta})$ is associated to the embedding of N in V.

In direct analogy with the results of 2.3 we have the following results on removable singularities for harmonic maps.

Sufficient Conditions for Removability

$f \in L^2_{1,loc}(M\setminus A,N)$, f Weakly Harmonic in $M\setminus A$

| | Condition on f | Condition on A |
|---|---|---|
| 4.1.1 | $f \in L^p_{1,loc}(M,N)$ | $B_{1,q}(A)=0$, $\frac{1}{p}+\frac{1}{q}=1$ <br> $2\leqslant p<\infty$ |
| 4.1.2 | $f \in C(M,N)$ | $B_{1,2}(A)=0$ |
| 4.1.3 | $f \in C^{\alpha}(M,N)\quad 0<\alpha\leqslant 1$ | $\lambda_{m-2+\alpha}(A)=0$ |
| 4.1.4 | $f \in C^1(M,N)$ | $\lambda_{m-1}(A)$ is locally finite |
| 4.1.5 | $f \in C^{1,\alpha}(M,N)$ | $\lambda_{m-1+\alpha}(A)=0$ |

The proofs of most of these results are essentially the same as for the linear case. A minor alteration is required by the presence of the quadratic term in (2); however, the hypotheses imply that this term is integrable, and that is all that is needed to handle this term. Proofs of 4.1.1 and 4.1.3-5 are contained in [EP]. The result of 4.1.2 is due to M. Meier ([ME1], [M2]). Meier has actually proved a stronger result: if f (M\A) is contained in a geodesic ball B(b,r) of center b and radius r, which does not meet the cut locus of b, and for which

$$r < \frac{\pi}{2\sqrt{B}}$$

where B > 0 is an upper bound for the sectional curvatures on B(b,r), then $f \in L_1^2(M,N)$. It follows that f is weakly harmonic in M. He also gives an example in which A is a point, $r = \frac{\pi}{2\sqrt{B}}$, and f does not have a weakly harmonic extension across A.

Since the results above are stated in terms of weak harmonicity, we remind the reader of the regularity result of Hildebrandt, Kaul, and Widman [HKW]: If $f \in L^2_{1,loc}(M,N)$ is weakly harmonic and f(M) is contained in a geodesic ball B(b,r) with the properties described in the previous paragraph, then f is infinitely differentiable and strongly harmonic.

4.1.6 It can be shown that the conditions in 4.1.1 with p > m and in 4.1.5 are necessary as well as sufficient.

Assume that M is a compact manifold, perhaps with a smooth boundary bM. Then we have

<u>Theorem</u>: Suppose $f_0 : M \to N$ is a harmonic map. Let $A \subset M$ be closed.

a) If $B_{1,q}(A) > 0$ $(\frac{1}{p} + \frac{1}{q} = 1,\ 2 \leq m < p < \infty)$, then there are maps $f \in L_1^p(M,N)$ arbitrarily close to $f_0$, such that f is harmonic on M\A, f is not harmonic on M,and $f = f_0$ on bM if $bM \neq \emptyset$.

b) If $\lambda_{m-1+\alpha}(A > 0$ $(0 < \alpha < 1)$, then there are maps $f \in C^{1,\alpha}(M,N)$ arbitrarily close to $f_0$, such that f is harmonic on $M \setminus A$, f is not harmonic on M, and $f = f_0$ on bM if $bM \neq \emptyset$.

For a proof see [EP].

4.1.7 No discussion of removable singularities for harmonic maps would be complete without stating the remarkable result of Sachs and Uhlenbeck ([SU]). Let $\Delta$ denote the unit disc in $\mathbb{C}$.

Theorem: Suppose $f \in L_1^2(\Delta,N)$ is harmonic in $\Delta \setminus \{0\}$. Then f is harmonic on $\Delta$.

By part a) of 4.1.1, f is weakly harmonic on $\Delta$. From this point of view the Sachs-Uhlenbeck result is a regularity result rather than a removable singularity result.

**4.2** **Another non-linear problem to which the results of the linear theory easily extend is of removable singularities for non-parametric minimal surfaces.**

Given a map $f: M \to N$, let $F: M \to M \times N$ be defined by $F(x) = (x,f(x))$. The condition that F be a minimal immersion takes the form

$$* \quad \begin{aligned} &\frac{\partial}{\partial x^i}\left(k^{ij}\sqrt{\det k}\right) = 0 \quad 1 \leq j \leq m \\ &\frac{\partial}{\partial x^i}\left(k^{ij}\sqrt{\det k}\, f_j^\alpha\right) = 0 \end{aligned}$$

where $k = g + f^*h$, (see [E], for example). We will consider the case where M is a domain in $\mathbb{R}^m$, and $N = \mathbb{R}^n$.

Sufficient Conditions for Removability
f Satisfies * Weakly on $M\setminus A$

| | Condition on f | Condition on A |
|---|---|---|
| 4.2.1 | None; n=1 | $\lambda_{m-1}(A) = 0$ |
| 4.2.2 | $f \in C^{0,1}(M)$ | $\lambda_{m-1}(A) = 0$ |
| 4.2.3 | $f \in C^{1}(M)$ | $\lambda_{m-1}(A)$ is locally finite |

The result in 4.2.1 is due to de Giorgi and Stampacchia [DGS] for the case when A is compact, with an earlier result due to Bers [BR]. The compactness restriction was removed by Leon Simon [SN]. The surprising fact here is that no growth restriction is required on f near the exceptional set A. The results in 4.2.2 and 4.2.3 are due to Harvey and Lawson [HL]. These results are proved in the same way as those of 2.3; the non-linearity in * causes no difficulty. The paper of Harvey and Lawson also contains important results about the extension of stationary and area minimizing rectifiable currents.

For completeness we mention the basic regularity result for weak solutions to *, due to Morrey [MO]: a weak $C^1$ solution to * is real analytic. Lawson and Osserman [LO] give examples of $C^{0,1}$ solutions to * which are not $C^1$.

Consider the map

$$f(x) = \begin{cases} 0 & x^m \leq 0 \\ x^m v & x^m \geq 0 \end{cases}$$

where $0 \neq v \in \mathbb{R}^n$. Then $f \in C^{0,1}(\mathbb{R}^m)$ and f satisfies * in $\mathbb{R}^m \setminus A$, where $A = \{x \mid x^m = 0\}$. This example illustrates many ways in which the hypothesis of 4.2 cannot be improved. In addition, if $A \subset M$ satisfies $\lambda_{m-1+\alpha}(A) > 0$, then the method used to prove 4.1.6 yields a map $f \in C^{1,\alpha}(M)$ which satisfies * in $M \setminus A$ but not in M.

## References


[AM] Adams, D. R., and Meyers, N. G., Bessel potentials. Inclusion relations among classes of exceptional sets. Ind. Univ.Math. J. 22 (1973), 873-905.

[AP] Adams, D. R., and Polking, J. C., The equivalence of two definitions of capacity. Proc. Am. Math. Soc. 37 (1973), 529-534.

[B1] Bochner, S., Weak solutions of linear partial differential equations. J. Math. Pures Appl., 35 (1956), 193-202.

[B2] Bochner, S., Partial differential equations and analytic continuations. PNAS, 38 (1952), 227-230.

[BE] Becker, J., Continuing analytic sets across $\mathbb{R}^n$. Math. Ann. 195 (1972), 103-106.

[BI] Bishop, E., Conditions for the analyticity of certain sets. Michigan Math. J.,11 (1964), 289-304.

[BR] Bers, L., Isolated singularities of minimal surfaces. Annals of Math. 53 (1951), 364-386.

[C1] Carleson, L., Removable singularities of continuous harmonic functions in $\mathbb{R}^m$. Math. Scand., 12 (1963), 15-18.

[C2] Carleson, L., Selected problems on exceptional sets. Van Nostrand, Princeton, 1967.

[CE] Cegrell, U., Sur les ensembles singuliers impropres des fonctions plurisousharmoniques, Compte Rendu, Acad. Sci. Paris 281 (1975), 905-908.

[CI] Cirka, E. M., On removable singularities of analytic sets. Soviet Math. Dok. 20 (1979).

[CLN] Chern, S. S., Levine, H. I., and Nirenberg, L., Intrinsic norms on a complex manifold in global analysis. Papers in Honor of K. Kodaira, Princeton Univ. Press, Princeton, NJ (1969) 119-139

[DGS] De Giorgi, E., and Stampacchia, G., Sulle singolarità eliminabili delle ipersuperfici minimali. Atti Acad. Naz. Lincei Rend. U. Sci. Fis. Mat. Nat. 38 (1965), 352-357.

[E] Eells, J., Minimal graphs. Manu. Math. 28 (1979), 101-108.

[EH] Ehrenpreis, L., A new proof of Hartog's theorem. BAMS 67 (1961), 507-509.

[EM] El Mir, M. H., Sur le prolongement des courants positifs fern These. Université Paris V (1982).

[EP] Eells, J., and Polking, J. C., Removable singularities of harmonic maps. (to appear in Indiana J. Math.).

[ES] Eells, J., and Sampson, J. H., Harmonic mappings of Riemannian manifolds. Am. J. Math. 86 (1964), 109-160.

[F] Federer, H., Geometric measure theory. Grundlehren Band 153, Springer-Verlag, New York (1969).

[FU] Funshashi, K., On the extension of analytic sets. Proc. Japan Acad. 54 (1978)

[GR] Grauert, H., and Remmert, R., Plurisubharmonische Funktionen in Komplexen Räumen. Math. Z., 65.

[H] Hörmander, L., An introduction to complex analysis in several variables. North Holland, Amsterdam (1973).

[H1] Harvey, R., Removable singularities for positive currents. Amer. J. Math. 96 (1974), 67-78.

[H2] Harvey, R., Removable singularities of cohomology classes in several complex variables. Amer. J. Math. 96 (1974), 498-504.

[HR] Hamilton, R. S., Harmonic maps of manifolds with boundary. Springer Lecture Notes No. 471, Springer-Verlag, New York (1975).

[HKW] Hildebrandt, S., Kaul, H., and Widman, K-O, An existence theorem for harmonic mappings of Riemannian manifolds. Acta Math. 138 (1977), 1-16.

[HL] Harvey, R., and Lawson, H. B., Extending minimal varieties. Inv. Math. 28 (1975), 209-226.

[HP1] Harvey, R., and Polking, J. C., Removable singularities of solutions of linear partial differential equations. Acta Math. 125 (1970), 39-56.

[HP2] Harvey, R., and Polking, J. C., A notion of capacity which characterizes removable singularities. T.A.M.S. 169 (1972), 183-195.

[HP3] Harvey, R., and Polking, J. C., Extending analytic objects. CPAM 28 (1975), 701-727.

[K] Kneser, M., Einige bemerkungen über das Minkowskische Flächenmass. Arch. Math. 6 (1955), 382-390.

[KI] King, J., The currents defined by analytic varieties. Acta Math. 127 (1971), 185-220.

[L1] Littman, W., Polar sets and removable singularities of partial differential equations. Ark. Math. 7 (1967), 1-9.

[L2] Littman, W., A connection between α capacity and m-p polarity. BAMS 73 (1967), 862-866.

[LO] Lawson, H. B., and Osserman, R., Non-existence, non-uniqueness, and irregularity of solutions to the minimal surface system. Acta Math. 139 (1977), 1-17.

[M] Meyers, N. G., A theory of capacities for potentials of functions in Lebesgue classes. Math. Scand. 26 (1970), 255-292.

[MA] Mazja, V. G., Removable singularities of bounded solutions of quasilinear elliptic equations of any order. J. Soviet Math. 3 (1975), 480–492.

[ME1] Meier, M., Removable singularities of bounded harmonic mappings and minimal submanifolds. Bonn Preprint No. 536 (1982).

[ME2] Meier, M., Removable singularities for weak solutions of quasilinear elliptic systems. J. Reine Angew. Math. 344 (1983), 87–101.

[MO] Morrey, C. B., Jr., Multiple integrals in the calculus of variations. Springer, Berlin–Hidelberg–New York (1966).

[R] Resetnjak, J. G., The concept of capacity in the theory of functions with generalized derivatives. Sibirsk Mat. Z. 10 (1969), 1109–1138.

[RS] Remmert, R., and Stein, K., Über die wesentlichen Singularitäten analytischer Mengen. Math. Ann. 126 (1953), 263–306.

[S1] Serrin, J., Removable singularities of solutions of elliptic equations. Arch. Rat. Mech. Anal. 17 (1964), 67–78.

[S2] Serrin, J., Local behavior of solutions of quasilinear equations. Acta Math. 111 (1964), 247–302.

[SE] Seeley, R. T., The C.I.M.E. conference of Stresa. Italy (1968).

[SH] Shiffman, B., On the removal of singularities of analytic sets. Michigan Math. J. 15 (1968), 111–120.

[SI] Siu, Y. T., Analyticity of sets associated to Lelong numbers and the extension of closed positive currents. Invent. Math. 27 (1974), 53–156.

[SK] Skoda, H., Prolongement des courants positifs fermés de masse finie. Inv. Math. 66 (1982), 361–376.

[SN] Simon, L., On a theorem of de Giorgi and Stampacchia. Math. Z. 155 (1977), 199–204.

[ST] Stoll, W., Uber die Fortsetzbarkeit analytischer Mengen endlichen Oberflacheninhaltes. Arch. Math. 9 (1958), 167–175.

[SU] Sachs, J., and Uhlenbeck, K., The existence of minimal immersions of 2-spheres. Ann. of Math. 113 (1981), 1–24.

[SY] Siboney, N., (to appear).

[T] Taylor, M. E., Pseudodifferential operators. Princeton University Press, Princeton, NJ (1981).

[V] Vladimirov, V. S., Methods of the theory of functions of many complex variables. Cambridge, Mass., The MIT Press (1966).

# Applications of the Maximum Principle

By M. H. Protter

Department of Mathematics
University of California
Berkeley, California 94720

## 1. Analogies with Analytic Functions

The connection between analytic functions of a complex variable and partial differential equations in the real domain is usually expressed in the observation that u and v, the real and imaginary parts of an analytic function $f = u + iv$, satisfy the Cauchy-Riemann equations

$$u_x - v_y = 0, \; v_x + u_y = 0.$$

The most powerful extension of the theory of analytic functions to first order systems of partial differential equations is the theory of pseudo-analytic functions developed by Bers [1] who showed that the basic elements of function theory carry over to solutions of systems of the form

$$u_x - v_y = a_1u + b_1v + c_1$$

$$v_x + u_y = a_2u + b_2v + c_2.$$

Except for fragmentary results, there has been no extension of complex function theory to solutions of partial differential equations or systems of equations in $R^n$ for $n > 2$. Nevertheless, there is a close analogy between certain basic theorems in analytic function theory and a corresponding set of theorems valid for solutions of second order elliptic partial differential equations in $R^n$ for all $n \geqslant 2$. Looked at with hindsight, we discern that these results in function theory acted as guideposts for the development of related results in differential equations. We shall see that in most cases the statements of the two results, one in function theory and the other in differential equations, have a compelling similarity. On the other hand, the proofs of the corresponding results are quite dissimilar since the rich structure of analytic function theory provides powerful techniques, while the meager structure of moderately smooth real-valued functions in $R^n$ forces the use of more complicated, involved methods. However, there is a reward for using the real variable methods in that the theorems originally developed as analogs to those in function theory have easily proven extensions; and yet these extensions do not have corresponding generalizations in function theory of the original results.

Our aim here is to exhibit several important theorems which show the relationship between complex function theory and second order linear elliptic equations and to show how these basic results were used to obtain extensions in the real case, ones which are not available in the complex case. We begin with the well-known maximum principle for analytic functions.

Theorem 1 (Maximum Modulus Principle). *Suppose that* f *is analytic in a bounded region* $\Omega \subset \mathbb{C}$, *and that* f *is continuous on* $\Omega \cup \partial\Omega$. *Let* $M = \max_{z \in \partial\Omega} |f(z)|$. *Then* $|f(z)| < M$ *for* $z \in \Omega$ *unless there is a point* $z_0 \in \Omega$ *for which* $|f(z_0)| = M$ *in which case* $f(z) \equiv Me^{i\alpha}$ *for some constant* $\alpha$.

The proof of the above theorem may be found, for example, in Titchmarsh [11].

Since the real and imaginary parts of analytic functions are harmonic, it is natural to expect that an analog to Theorem 1 is valid for solutions of the Laplace equation. Indeed, it has been known at least since the time of Gauss that harmonic functions satisfy a maximum principle. However, it was not until 1927 that E. Hopf [2] established a result corresponding to Theorem 1 for solutions of general second order linear elliptic equations.

We define the operator

$$(1) \qquad Lu \equiv \sum_{i,j=1}^{n} a_{ij}(x) \frac{\partial^2 u}{\partial x_i \, \partial x_j} + \sum_{i=1}^{n} b_i(x) \frac{\partial u}{\partial x_i}$$

for x in a domain $\Omega$ in $\mathbb{R}^n$. Then L is said to be *elliptic at a point* $x \in \Omega$ if there is a positive number $\mu$ such that

$$(2) \qquad \sum_{i,j=1}^{n} a_{ij}(x)\xi_i \xi_j \geqslant \mu \sum_{i=1}^{n} \xi_i^2$$

for all $\xi = (\xi_1,\ldots,\xi_n)$ in $\mathbb{R}^n$. The operator is uniformly elliptic in $\Omega$ if there is a number $\mu_0 > 0$ such that (2) holds with $\mu \geqslant \mu_0$ for all $x \in \Omega$. The basic maximum principle is contained in the next result.

Theorem 2 (Strong Maximum Principle of Hopf). *Let* u *satisfy in* $\Omega$ *the differential inequality* $Lu \geqslant 0$ *where* L, *given by* (1), *is uniformly elliptic. Suppose the coefficients* $a_{ij}$, $b_i$, $i,j = 1,2,\ldots,n$ *are uniformly bounded in* $\Omega$. *If* u *attains a maximum value* M *at any point of* $\Omega$, *then* $u \equiv M$ *in* $\Omega$.

We observe that it is sufficient for u to satisfy a differential inequality in order to derive the maximum principle. If u is a solution of $Lu \leqslant 0$ in $\Omega$, there is a corresponding *strong minimum principle* which states that if the minimum m of u is achieved at some point of $\Omega$, then $u \equiv m$ in $\Omega$. For solutions of $Lu = 0$ in $\Omega$ there is both a maximum and a minimum principle, and hence the maximum modulus theorem corresponding to Theorem 1 holds for all solutions of linear elliptic second order equations $Lu = 0$.

The *weak maximum principle* asserts that for a bounded solution u of $Lu \geqslant 0$ in a bounded region $\Omega$, the maximum of u must be achieved on the boundary $\partial\Omega$ and may also be achieved at interior points of $\Omega$. Theorems 1 and 2 are both much stronger in that they state that nonconstant functions cannot achieve a maximum at any interior point of $\Omega$.

An important extension of Theorem 2 is the *Boundary point maximum principle*, also obtained by Hopf [3]. Suppose that u is a nonconstant solution of $Lu \geqslant 0$ in $\Omega$ and that the maximum M of u is achieved at a boundary point $P \in \partial\Omega$. Suppose also that $\frac{\partial u}{\partial \nu}$ exists at P where $\frac{\partial}{\partial \nu}$ is any directional derivative "pointing outward" from $\Omega$. Since the maximum of u occurs at P, it is clear that $\frac{\partial u(P)}{\partial \nu} \geqslant 0$, for otherwise u would have a value larger than M at some point in $\Omega$ near P. The remarkable result of Hopf states that

$$\frac{\partial u(P)}{\partial \nu} > 0$$

i.e., any outward directional derivative at P is strictly positive. A corresponding boundary point minimum principle holds for solutions of $Lu \leqslant 0$ in $\Omega$. The boundary principles have many important applications, especially in the development of a priori bounds for solutions of both linear and quasilinear second order elliptic partial differential equations. It is not difficult to show that the Strong Maximum Principle follows from the Boundary Point Principle. Since there is no counterpart to the boundary result in complex function theory, we have an example of an extension in the real variable case which has its roots in a more restricted result for complex functions.

Two additional extensions of the maximum principle are noteworthy. With L given by (1), suppose that u is a solution in $\Omega$ of the differential inequality

$$(L + h)u \geqslant 0$$

where h(x) is a nonpositive bounded function in $\Omega$. Then u satisfies the modified strong maximum principle: *if the maximum* M *of* u *occurs at a point of* $\Omega$ *and* $M \geqslant 0$, *then* $u \equiv M$ *in* $\Omega$. It is easy to find counterexamples if $M < 0$ or if h is not nonpositive throughout $\Omega$.

A further extension of the maximum principle allows us to draw the appropriate strong result when h is either positive or negative.

Theorem 3 (Generalized Strong Maximum Principle). *Let* u *satisfy in* $\Omega$ *the differential inequality* $(L + h)u \geqslant 0$. *If there exists a function* w *defined in* $\Omega \cup \partial\Omega$ *such that*

(i) $w(x) > 0$ *for* $x \in \Omega \cup \partial\Omega$,

(ii) $(L + h)w \leqslant 0$ *in* $\Omega$,

*then* $u/w$ *cannot attain a nonnegative maximum in* $\Omega$ *unless* $u/w$ *is constant.*

We observe that no hypothesis on the sign of $h$ is made (although $h$ must be bounded). Also, there is an analogous Boundary Maximum Principle: *if* $u(x)/w(x)$ *attains its nonnegative maximum at a boundary point* $P$, *and* $u/w$ *is not constant, then*

$$\frac{\partial}{\partial \nu}\left(\frac{u}{w}\right) > 0 \text{ at } P.$$

It is natural to ask whether or not there always exist functions $w$ satisfying conditions (i) and (ii) of Theorem 3. It can be shown that if $\Omega$ is sufficiently small in some sense (e.g., the diameter of $\Omega$ is small), then a function $w$ as in Theorem 3 can always be found explicitly. On the other hand, if $h(x)$ is positive and if $\Omega$ is large, it can be shown that no function $w$ exists which satisfies both (i) and (ii). The size and shape of $\Omega$ needed to get an appropriate $w$ depend only on the coefficients of $L$ and on $h$. (See [8], p. 133.)

The Hadamard three-circles theorem is one of the beautiful elementary results in analytic function theory, one which is a direct consequence of the Maximum Modulus Principle.

Theorem 4 (Hadamard Three-Circles Theorem). *Let* $K_1, K, K_2$ *be three concentric disks of radii* $R_1 < r < R_2$, *respectively. Suppose* $f$ *is analytic in a region containing* $\bar{K}_2 - K_1$, *and denote by* $M_1$, $M(r)$, $M_2$ *the maxima of* $|f(z)|$ *on* $\partial K_1$, $\partial K$, $\partial K_2$, *respectively. Then*

$$[M(r)]^{\log(R_2/R_1)} \leqslant M_1^{\log(R_2/r)} M_2^{\log(r/R_1)}$$

*Equality occurs only if* $f(z)$ *is a constant multiple of a power of* $z$.

Since the logarithm is an increasing function, we may write the above inequality as

$$(3) \qquad \log M(r) \leq \frac{\log R_2 - \log r}{\log R_2 - \log R_1} \log M_1 + \frac{\log r - \log R_1}{\log R_2 - \log R_1} \log M_2.$$

Inequality (3) asserts that for $R_1 < r < R_2$, the function $\log M(r)$ is a convex function $\log r$.

The three-circles theorem has several analogs for elliptic equations in $R^n$. The simplest one applies to subharmonic functions (i.e., functions satisfying the differential inequality $\Delta u \geq 0$) and is known as the three-spheres theorem.

Theorem 5 (Three-Spheres Theorem). *Let* $K_1, K, K_2$ *be three concentric balls in* $R^n$, $n \geq 2$ *of radii* $R_1 < r < R_2$, *respectively. Suppose* $u$ *is a solution of* $\Delta u \geq 0$ *in a region containing* $\bar{K}_2 - K_1$, *and denote by* $M_1$, $M(r)$, $M_2$ *the maxima of* $u(x)$ *on* $\partial K_1$, $\partial K$, $\partial K_2$, *respectively. Then*

$$(4a) \qquad M(r) \leq \frac{\log R_2 - \log r}{\log R_2 - \log R_1} M_1 + \frac{\log r - \log R_1}{\log R_2 - \log R_1} M_2$$

if $n = 2$

$$(4b) \qquad M(r) \leq \frac{r^{2-n} - R_2^{2-n}}{R_1^{2-n} - R_2^{2-n}} M_1 + \frac{R_1^{2-n} - r^{2-n}}{R_1^{2-n} - R_2^{2-n}} M_2 \text{ if } n \geq 3.$$

Inequality (4a) states that $M(r)$ is a convex function of $\log r$, while (4b) states that $M(r)$ is a convex function of $r^{2-n}$. We note that Theorem 5 is not quite the analog of Theorem 4 in that $M(r)$ rather than $\log M(r)$ is shown to have a convexity property. Inequalities related to (4a), (4b) are valid for the minimum value on concentric spheres of solutions of $\Delta u \leq 0$ (superharmonic functions).

We now describe two extensions of Theorem 5. Suppose first that $u$ is a solution of $Lu \geq 0$ in the region described in Theorem 5, with $L$ given by (1). From the maximum principle, it follows that if $u$ is a nonconstant solution between $K_1$ and $K_2$, then $M(r)$ cannot be constant in any interval. In fact, $M(r)$ can have no maximum and, hence, at most one minimum for $R_1 < r < R_2$; thus either $M(r)$ is always decreasing, always increasing, or it may first decrease and then increase. For simplicity, suppose that $M(r)$ is an increasing function for $R_1 < r < R_2$, although the following theorem holds for all three types of behavior of $M(r)$.

Theorem 6. *Suppose* $K_1, K, K_2$ *are as in Theorem* 5 *and that there exists a function* $\psi(r)$, *increasing for* $R_1 < r < R_2$, *and such that* $L\psi \leq 0$ *in* $\bar{K}_2 - K_1$. *Let* u *be a solution of* $Lu \geq 0$ *in a region containing* $\bar{K}_2 - K_1$, *and denote by* $M_1, M(r), M_2$ *the maxima of* u *on* $\partial K_1$, $\partial K$, $\partial K_2$, *respectively. If* $M(r)$ *is an increasing function of* r, *then*

$$M(r) \leq \frac{\psi(R_2) - \psi(r)}{\psi(R_2) - \psi(R_1)} M_1 + \frac{\psi(r) - \psi(R_1)}{\psi(R_2) - \psi(R_1)} M_2. \tag{5}$$

Inequality (5) states that $M(r)$ is a convex function of $\psi(r)$. Since the functions $\psi(r) = \log r$ for $n = 2$ and $\psi(r) = r^{2-n}$ for $n \geq 3$ satisfy the conditions of Theorem 6 with $L = \Delta$, formulas (4a) and (4b) are special cases of (5). More generally, if L is uniformly elliptic and the coefficients of L are uniformly bounded, it is always possible to find functions $\psi$ which satisfy the hypotheses of Theorem 6.

A second extension of Theorem 5 treats the case of surfaces in $R^n$ other than spheres. Suppose that for $R_1 < r < R_2$ and for $x \in R^n$, the equation

$$f(x) = r$$

represents a one-parameter family $S_r$ of smooth closed (n-1)-dimensional hypersurfaces contained in a domain $\Omega$ in which $Lu \geq 0$. Once again it is possible to define a function $\psi(r)$ in $\Omega$ so that

$$M(r) = \max_{x \in S_r} u(x)$$

is a convex function of $\psi(r)$. For example, it is a simple matter to establish a convexity theorem for the maximum values of a subharmonic function on a one-parameter family of confocal ellipses in the plane.

Liouville's theorem states: *a function which is analytic for all finite values of* $z \in \mathbb{C}$, *and is bounded must be a constant.* Attempts at establishing the counterpart of this result for linear and nonlinear elliptic equations and systems have led to surprisingly rich results, as well as to a number of conjectures and counterexamples. In the simplest analogous case for real-valued functions, we show that a Liouville-type result holds for subharmonic functions in a plane under conditions much less restrictive than in the complex case.

<u>Theorem</u> <u>7</u>. *Suppose that* u *is subharmonic in the whole plane except possibly at one point, taken to be the origin. Let* $K_r$ *denote the disk of radius* r, *center at the origin, and let* M(r) *be the maximum of* u *on* $\partial K_r$. *Assume that as* $r \to \infty$ *and* $r \to 0$,

$$\liminf \frac{M(r)}{|\log r|} \leqslant 0.$$

*Then* u *is a constant.*

The proof is a direct consequence of the three-spheres case for n = 2 (formula (4a)). We let $R_2 \to \infty$ in that formula and deduce that

$$M(r) \leqslant M(R_1) \text{ for } r \geqslant R_1.$$

Similarly, letting $R_1 \to 0$, we obtain

$$M(r) \leqslant M(R_2) \text{ for } r \leqslant R_2.$$

Since $R_1$ and $R_2$ are arbitrary, we conclude that M(r) is constant. By the Strong Maximum Principle, $u \equiv$ constant.

In contrast with the Liouville Theorem for complex functions, Theorem 7 does not require u to be bounded. If u has slower than logarithmic growth from above, then u is constant. In addition, there is no requirement on the behavior of u from below so that u may tend to $-\infty$ at any speed as $r \to \infty$ or 0. Finally, a single point can always be excluded from the region in which $\Delta u \geq 0$ so long as the growth of u near that point is less than logarithmic.

Theorem 7 is false in $R^n$, $n \geq 3$. It is not difficult to exhibit a bounded subharmonic function in all of $R^3$ which is not constant. However, it is true that a bounded harmonic function defined in all of $R^n$ is always constant. Moreover, the result is extendable to entire solutions of general linear uniformly elliptic second order equations. The techniques generally require estimates for the gradient of a solution as does the proof of the complex Liouville theorem. The methods for obtaining these gradient bounds are quite intricate in the real case while in the complex case they result directly from the Cauchy formula.

The Phragmén-Lindelöf Theorem for analytic functions describes the behavior of an analytic function near a point on the boundary of the domain of analyticity. The main result, an important extension of the Maximum Modulus Principle, shows that analytic functions must remain bounded in a neighborhood of the boundary point under rather mild hypotheses on the growth of the function. For simplicity, we state the result for a sector with the point at infinity as the boundary point.

Theorem 8 (Phragmén-Lindelöf). *Let* $\Omega$ *be a sector in the plane bounded by the lines* $\theta = \pm \frac{\pi}{2\alpha}$, *where* $r, \theta$ *are polar coordinates and* $\alpha > 0$ *is a constant. Let* f *be an analytic function in a region containing* $\bar{\Omega}$ *and suppose that*

$$|f(z)| \leq M \textit{ on the lines } \theta = \pm \frac{\pi}{2\alpha}, \quad z = re^{i\theta}.$$

*If, for some* $\beta < \alpha$

$$f(z) = O(\exp r^{\beta}) \textit{ as } r \to \infty$$

*then* $|f(z)| \leqslant M$ *in* $\Omega$.

The direct counterpart of Theorem 8 for subharmonic functions in the plane is less restrictive than the classical Phragmén-Lindelöf result.

Theorem 9 (Phragmén-Lindelöf). *Suppose* $\Delta u \geqslant 0$ *in the sector described in Theorem* 8 *and that* $u \leqslant M$ *on the rays* $\theta = \pm\pi/2\alpha$. *Assume that*

$$\liminf_{R\to\infty} \{R^{-\alpha} \max_{r=R} u(r,\theta)\} \leqslant 0.$$

*Then* $u \leqslant M$ *in* $\Omega$.

The proof of Theorem 9 is a consequence of the Strong Maximum Principle applied to a differential inequality which has the Laplacian as its principal part. The details are given in [8].

As was the case in Liouville's Theorem, the hypotheses of the real ariable Phragmén-Lundelöf principle are much less restrictive than those of the complex variable principle. In the real case, u is merely required to satisfy a differential inequality, there are no requirements on the behavior of u from below, and the growth condition of u in Theorem 9 is milder than that of f in Theorem 8. Moreover, the result is applicable to general differential inequalities in $R^n$ and to general domains. The underlying result states that if the solution to a linear second order elliptic differential inequality does not grow too rapidly in the neighborhood of a boundary point, then it does not grow at all in this neighborhood, i.e., the solution remains bounded from above near that point.

## 2. Gradient Bounds

The maximum principle provides a bound for solutions of linear elliptic second order equations in any region in terms of the maximum of the boundary values of the solution. In establishing existence theory for solutions to boundary value problems for such equations, it is necessary to obtain a priori bounds, not only for the solution itself but also for its derivatives. Since the derivatives do not necessarily satisfy an elliptic differential equation, or even an elliptic differential inequality, there is no direct way to apply the maximum principle to obtain estimates for the gradient of a solution in terms of its boundary values. Nevertheless, there is an indirect technique for achieving this desired result.

We consider the elliptic equation

$$\text{(6)} \qquad \sum_{i,j=1}^{n} a_{ij}(x) \frac{\partial^2 u}{\partial x_i \partial x_j} + \sum_{i=1}^{n} b_i(x) \frac{\partial u}{\partial x_i} - h(x)u = 0$$

and we define the quantity

$$\text{(7)} \qquad w(x) = \sum_{i=1}^{n} \left(\frac{\partial u}{\partial x_i}\right)^2 + \gamma u^2.$$

It is possible to show that for a sufficiently large value of the constant $\gamma$, the function w satisfies the maximum principle. More precisely, we have the next result (see [7]).

<u>Theorem</u> <u>10</u>. *Let* u *be a* $C^3$ *solution of* (6) *in a bounded region* $\Omega$, *and suppose that* $a_{ij}$, $b_i$ *and* h *are* $C^1$ *functions in* $\Omega$ *with* $h \geq 0$. *Then there is a positive number* $\gamma$ *whose size depends only on the coefficients in* (6) *such that the function* w *given by* (7) *satisfies the maximum principle in* $\Omega$.

Once the constant $\gamma$ is determined, it is clear that the gradient at any point of a domain $\Omega$ is bounded by a combination of the gradient on $\partial\Omega$ and the maximum of u on $\partial\Omega$. However, it is possible to achieve a more useful inequality, one which provides an estimate for the gradient at any interior point of $\Omega$ in terms of the maximum of the solution on the boundary. This results from the following extension [7] of Theorem 10. Let $\alpha$ be an arbitrary $C^2$ function on $\bar{\Omega}$. Then under the same hypotheses as given in Theorem 10, there is a constant $\gamma$ such that

$$v(x) = [\alpha(x)]^2 \sum_{i=1}^{n} \left[\frac{\partial u}{\partial x_i}\right]^2 + \gamma u^2$$

satisfies the maximum principle in $\Omega$. Of course, the number $\gamma$ depends on $\alpha$ as well as the coefficients of the equation. In particular, suppose that $\alpha$ is chosen to be positive in $\Omega$ and zero on $\partial\Omega$. Since the maximum of v occurs on $\partial\Omega$, we see that at any point $x \in \Omega$, the gradient is bounded by a constant times the maximum of u on $\partial\Omega$. Since $\alpha(x) \to 0$ as $x \to \partial\Omega$, this bound is uniform only on compact subsets of $\Omega$. Thus, we get an interior gradient bound of the type required for existence theory.

The maximum principle may also be employed for gradient estimates of quasilinear elliptic equations and systems. See Ladyzenskaya and Uraltseva [4] and Serrin [9]. Payne and Stakgold [5] employed a technique quite different from that in [4,9], one which is a natural generalization of Theorem 10. We describe it in the case of the semilinear equation

(8) $\quad \Delta u + f(u) = 0,$

where f is a nonnegative $C^1$ function of its argument. Define the function

(9) $\quad F(x) = |\nabla u|^2 + 2 \int_0^u f(t)dt,$

where $\nabla u$ denotes the gradient of u. A computation yields

$$\Delta F \equiv \sum_{k=1}^{n} \frac{\partial^2 F}{\partial x_k^2} = 2 \sum_{i,k=1}^{n} \left\{\frac{\partial^2 u}{\partial x_i \partial x_k}\right\}^2 - 2f'(u)\, |\nabla u|^2 - 2f^2(u)$$

$$+ 2f'(u)\, |\nabla u|^2 = 2 \sum_{i,k=1}^{n} \left[\frac{\partial^2 u}{\partial x_i \partial x_k}\right]^2 - 2f^2(u),$$

in which a simplification has been made by differentiating the differential equation (8) and eliminating third derivatives. A further computation shows that

$$(10) \qquad \Delta F - \frac{1}{2|\nabla u|^2} \sum_{k=1}^{n} \left\{\frac{\partial F}{\partial x_k} - 4f(u) \frac{\partial u}{\partial x_k}\right\} \frac{\partial F}{\partial x_k} \geqslant 0$$

Thus, we see that F satisfies an elliptic differential inequality and, hence, the maximum principle. However, the basic theorem of Hopf requires that the coefficients in the differential inequality (10) remain bounded. Therefore, we may apply the principle to F only at points where $|\nabla u| \neq 0$, since this quantity appears in the denominator of the coefficient of $\partial F/\partial x_k$. Nevertheless, we obtain a gradient bound in all cases. For either F assumes its maximum on the boundary or F assumes its maximum at a point where $|\nabla u| = 0$. In the latter case, F is bounded by $\int_0^{u_M} f(t)dt$ where $u_M$ is the maximum of u in $\bar{\Omega}$. In the former case, F is bounded by a combination of the maxima of $|\nabla u|$ and u on the boundary.

The result for equation (8) has been extended to operators in which $\Delta u$ is replaced by a general second order linear elliptic operator [6]. Further extensions in a variety of directions may be found in the book by Sperb [10].

## Bibliography


[1] Bers, L., Theory of Pseudoanalytic Functions. Lecture Notes, New York University (1953).

[2] Hopf, E., Elementare Bemerkungen über die Lösengen partielles Differentialgleichungen zweiter Ordnung vom Elliptischen Typus.

[3] Hopf, E., A Remark on Linear Elliptic Differential Equations of the Second Order. Proc. Amer. Math. Soc., Vol. 3 (1952), 791-793.

[4] Ladyzenskaya, O. A., and Uraltseva, N. N., Local Estimates for Gradients of Solutions of Non-Uniformly Elliptic and Parabolic Equations. Comm. Pure and Appl. Math., Vol. 23 (1970), 677-703.

[5] Payne, L., and Stakgold, I., On the Mean Value of the Fundamental Mode in the Fixed Membrane Problem. J. Appl. Analysis, Vol. 3 (1973), 295-306.

[6] Protter, M. H., The Maximum Principle and Eigenvalue Problems. Proceedings of the 1980 Beijing Symposium on Differential Geometry and Differential Equations.

[7] Protter, M. H., and Weinberger, H. F., A Maximum Principle and Gradient Bounds for Linear Elliptic Equations. Indiana Univ. Math. Journal, Vol. 23 (1973), 239-249.

[8] Protter, M. H., and Weinberger, H. F., Maximum Principles in Differential Equations. Prentice-Hall, Inc. (1967).

[9] Serrin, J., Gradient Estimates for Solutions of Nonlinear Elliptic and Parabolic Equations. Symposium on Nonlinear Functional Analysis, Univ. of Wisconsin (1972).

[10] Sperb, R., Maximum Principles and Their Applications. Academic Press (1981).

[11] Titchmarsh, E. C., The Theory of Functions. Oxford Univ. Press (1932).

# Minimax Methods and Their Application to Partial Differential Equations

By Paul H. Rabinowitz
Department of Mathematics
University of Wisconsin
Madison, Wisconsin 53706

This research was sponsored in part by the National Science Foundation under Grant No. MCS-8110556. Reproduction in whole or in part is permitted for any purpose of the United States Government.

Minimax methods are a useful tool in obtaining critical points for real valued functionals defined on a Banach or Hilbert space or manifold. In this lecture we will discuss the use of such methods and show how they can be applied to obtain existence theorems for semilinear elliptic boundary value problems. There has been a considerable amount of work in this area during the past several years and we will only be able to touch on a small number of topics. For more extensive views of this and related fields, see Nirenberg [1], Rabinowitz [2], [3], Berger [4], Palais [5], Schwartz [6], Krasnoselski [7], Vainberg [8].

To begin, let E be a real Banach space and I a continuously differentiable real valued functional on E. A critical point of I is a point $u_0 \in E$ at which the Fréchet derivative of I vanishes, i.e., $I'(u_0) = 0$. We then call $I(u_0)$ a critical value of I. Since $I'(u_0)$ is a linear map from E to $\mathbb{R}$, the vanishing of $I'(u_0)$ means

(1) $$I'(u_0)\varphi = 0$$

for all $\varphi \in E$. Equation (1) provides a connection between critical points of real valued functionals and weak solutions of differential equations. E.g., consider the semilinear elliptic boundary value problem:

(2) $$\begin{aligned} -\Delta u &= p(x,u), \quad x \in \Omega \\ u &= 0, \quad\quad\quad x \in \partial\Omega \end{aligned}$$

where $\Omega$ is a bounded domain in $\mathbb{R}^n$ with a smooth boundary and p is a given function which is say smooth. We will call a function u satisfying

(3) $$\int_\Omega [\nabla u \cdot \nabla\varphi - p(x,u)\varphi]\, dx = 0$$

for all $\varphi \in C_0^\infty(\Omega)$ a weak solution of (2). Note that (3) is obtained from (2) by multiplying by $\varphi$ and integrating by parts. Thus, satisfying

(3) is less restrictive then satisfying (2). However, if, e.g., u belongs to the Sobolev space $W_0^{1,2}(\Omega)$--the closure of $C_0^\infty(\Omega)$ with respect to

$$\|u\|^2_{W_0^{1,2}(\Omega)} = \int_\Omega |\nabla u|^2 dx \equiv \|u\|^2$$

--and p is locally Hölder continuous and satisfies an appropriate growth condition, it is known that any weak solution of (2) is also a classical solution. Thus, to solve (2), it suffices to find solutions of (3).

Let $P(x,\xi)$ denote the primitive of $p(x,\xi)$, i.e.,

$$P(x,\xi) = \int_0^\xi p(x,t)\, dt.$$

Further suppose

(4) $$I(u) = \int_\Omega \left[\frac{1}{2} |\nabla u|^2 - P(x,u)\right] dx$$

belongs to $C^1(E,\mathbb{R})$ for $u \in E \equiv W_0^{1,2}(\Omega)$. (Indeed, this will be the case under the hypotheses on p mentioned above which imply weak solutions of (2) are classical solutions.) Then

(5) $$I'(u)\varphi = \int_\Omega [\nabla u \cdot \nabla \varphi - p(x,u)\varphi]\, dx$$

and at a critical point $u_0$ of I we see from (3)-(5) that $u_0$ is a weak solution of (2).

This observation allows us to use ideas and methods from critical point theory to find (weak) solutions of partial differential equations which possess a variational formulation. Some of the basic ideas for the abstract results we describe go back to early work of Ljusternik and Schnirelmann [9]. One can also hope to use Morse theory, but this generally requires $C^2$ functionals with nondegenerate critical points rather than the $C^1$ case we shall treat and, therefore, has been less useful in applications to partial differential equations up to now.

How does one find critical points of functionals $I \in C^1(E,\mathbb{R})$? For $E = \mathbb{R}^n$, we know one can hope to find global or possibly local maxima or minima of I. However, there may not exist any such points. In fact, in this lecture our main focus will be on <u>indefinite functionals</u>. Without offering a precise definition, what we have in mind here are functionals which are not bounded from above or below even modulo subspaces or submanifolds of finite dimension or codimension. As a simple example, consider a one-dimensional version of (2):

$$(6) \qquad \begin{aligned} -u'' &= u^3, \quad 0 < x < 1 \\ u(0) &= 0 = u(1). \end{aligned}$$

Solutions of (6) correspond to critical points of

$$I(u) = \int_0^1 \left[\frac{1}{2} |u'|^2 - \frac{1}{4}u^4\right] dx$$

Note that for $u \not\equiv 0$ and $\alpha \in \mathbb{R}$. $I(\alpha u) \to -\infty$ as $|\alpha| \to \infty$. Consequently, I is not bounded from below. Further, choosing $u_k = \sin k\pi x$, we see $I(u_k) \to \infty$ as $k \to \infty$. Thus, I also is not bounded from above and is an indefinite functional.

In order to find critical points of a functional, one generally needs some "compactness" for the problem. A convenient way to guarantee this is to assume that I satisfies the <u>Palais-Smale condition</u> (PS): Every sequence $(u_m)$ such that $I(u_m)$ is bounded and $I'(u_m) \to 0$ has a convergent subsequence.

Now we can describe our first abstract critical point theorem, the so-called Mountain Pass Theorem which is due to Ambrosetti and the author [1-3], [10]. Let $B_\rho$ denote the open ball of radius $\rho$ about 0 in E and $\partial B_\rho$ its boundary.

<u>Theorem 7</u>: Let E be a real Banach space and $I \in C^1(E,\mathbb{R})$, I satisfying (PS). Suppose $I(0) = 0$,

$(I_1)$ There exists $\alpha,\rho > 0$ such that $I|_{\partial B_\rho} \geq \alpha$

and

$(I_2)$ There exists $e \in E\setminus B_\rho$ such that $I(e) \leq 0$.

Then $I$ possesses a critical value $c \geq \alpha$. Moreover, $c$ can be characterized as

(8) $$c = \inf_{g \in \Gamma} \max_{t \in [0,1]} I(g(t))$$

where $\Gamma = \{g \in C([0,1],E) \mid g(0) = 0, g(1) = e\}$.

<u>Remark</u>: Geometrically the Mountain Pass Theorem says if 0 and $e$ are separated by a mountain range, there exists a mountain pass--the desired critical point--somewhere "between" them.

Since it illustrates in a simple setting the main ideas used in minimax theory, we will sketch the proof of the Mountain Pass Theorem. The basic ingredients are a "Deformation Theorem" followed by an indirect minimax argument. For $s \in \mathbb{R}$, set $A_s = \{u \in E \mid I(u) \leq s\}$ and $K_s = \{u \in E \mid I(u) = s \text{ and } I'(u) = 0\}$.

<u>Theorem 9 (Deformation Theorem)</u>: Let $E$ be a real Banach space and $I \in C^1(E,\mathbb{R})$ satisfying (PS). If $c \in \mathbb{R}$, $\bar{\varepsilon} > 0$, and $K_c = \emptyset$, there exists $\varepsilon \in (0,\bar{\varepsilon})$ and $\eta \in C([0,1] \times E,E)$ such that

1° $\eta(t,u) = u$ if $I(u) \notin [c - \bar{\varepsilon}, c + \bar{\varepsilon}]$

and

2° $\eta(1,A_{c+\varepsilon}) \subset A_{c-\varepsilon}$.

If $E = \mathbb{R}^n$ and $I \in C^2$, $\eta$ is obtained by taking a suitably localized negative gradient flow corresponding to $I$, i.e., the solution to an ordinary differential equation of the form

$$\frac{d\eta}{dt} = -\psi(\eta)I'(\eta)$$

$$\eta(0,u) = u.$$

The general case uses similar ideas but with considerably more technicalities involved. See, e.g., [3], [5], or [11] for details.

<u>Proof of the Mountain Pass Theorem</u>: Since each curve $g \in \Gamma$ crosses $\partial B_\rho$, by $(I_1)$,

$$\max_{t \in [0,1]} I(g(t)) \geq \alpha.$$

Therefore, $c \geq \alpha$ by (8). Suppose $c$ is not a critical value of I. Then by the Deformation Theorem with $\bar{\varepsilon} = \frac{\alpha}{2}$, there exists $\varepsilon \in (0,\frac{\alpha}{2})$ and $\eta \in C([0,1] \times E,E)$ such that $\eta$ satisfies 1°, 2° above. Choose $g \in \Gamma$ such that

$$(10) \qquad \max_{t \in [0,1]} I(g(t)) \leq c + \varepsilon.$$

Consider $h(t) \equiv \eta(1,g(t))$. Clearly, $h \in C([0,1],E)$. Since $g(0) = 0$, $g(1) = e$, and $I(0) = 0 \geq I(e)$, by 1° and our choice of $\bar{\varepsilon}$, $h(0) = \eta(1,0) = 0$ and $h(1) = \eta(1,e) = e$. Therefore, $h \in \Gamma$ so

$$(11) \qquad \max_{t \in [0,1]} I(h(t)) \geq c.$$

But $g([0,1]) \subset A_{c+\varepsilon}$ via (10) which implies

$$(12) \qquad \max_{t \subset [0,1]} I(h(t)) \leq c - \varepsilon$$

by 2° of the Deformation Theorem. Comparing (11) and (12) yields a contradiction and the proof is complete.

Next we sketch two applications of Theorem 7. Consider, first, (2) where p satisfies:

($p_1$) $p(x,\xi)$ is locally Lipschitz continuous.

($p_2$) $|p(x,\xi)| \leq a_1 + a_2|\xi|^s$ where $1 \leq s < \frac{n+2}{n-2}$.

($p_3$) $p(x,\xi) = o(|\xi|)$ as $\xi \to 0$.

($p_4$) $0 < \mu P(x,\xi) \leq \xi p(x,\xi)$ for large $|\xi|$ where $\mu > 2$.

<u>Remark</u> <u>13</u>: (i) These hypotheses are satisfied if, e.g., $p(x,\xi) = |\xi|^{s-1}\xi$. (ii) Hypothesis ($p_2$) can be dropped if $n = 1$ and weakened if $n = 2$. (iii) ($p_1$) and ($p_2$) ensure that a weak solution of (2) is a classical solution.

Letting $E = W_0^{1,2}(\Omega)$ and defining $I(u)$ by (4), hypotheses $(p_1)$ and $(p_2)$ imply that $I \in C^1(E,\mathbb{R})$. Clearly, $I(0) = 0$. Condition $(p_3)$ implies $P(x,\xi) = o(|\xi|^2)$ as $\xi \to 0$. Simple estimates then show

$$\int_\Omega P(x,u)\,dx = o(\|u\|^2)$$

as $u \to 0$ in E. Since

$$I(u) = \frac{1}{2}\|u\|^2 - \int_\Omega P(x,u)dx = \frac{1}{2}\|u\|^2 + o(\|u\|^2)$$

as $u \to 0$, it then follows that I satisfies $(I_1)$ of Theorem 7. By $(p_4)$, there exist constants $a_3$, $a_4 \geq 0$ such that for all $\xi \in \mathbb{R}$,

$$(14) \qquad P(x,\xi) \geq a_3|\xi|^\mu - a_4.$$

Therefore, for $u \in \partial B_1$,

$$I(\alpha u) \leq \frac{1}{2}\alpha^2 - \alpha^\mu a_3 \int_\Omega |u|^\mu dx + a_4 \text{ meas } \Omega \to -\infty$$

as $\alpha \to \infty$. Hence, $(I_2)$ is satisfied. Finally, it is not difficult to use $(p_4)$ and the form of I to verify that I satisfies (PS). See, e.g., [10], [2], or [3] for more details. Thus, by the Mountain Pass Theorem we have

<u>Theorem</u> <u>15</u>: If p satisfies $(p_1)$ - $(p_4)$, (2) possesses a weak solution $\bar{u}$ such that $I(\bar{u}) \geq \alpha > 0$.

<u>Remark</u> <u>16</u>: By $(p_3)$, $u \equiv 0$ is a solution of (2). Since $I(\bar{u}) \geq \alpha > 0 = I(0)$, $\bar{u}$ is a nontrivial weak solution of (2). Further arguments based on the maximum principle show (2) has a positive and a negative solution in $\Omega$. See [10], [2], or [3].

For our second application, consider the nonlinear eigenvalue problem

$$(17)\qquad \begin{aligned} -\Delta u &= \lambda p(u), \quad x \in \Omega \\ u &= 0, \qquad x \in \partial\Omega \end{aligned}$$

where $\Omega$ is as earlier, $\lambda \geq 0$, and p satisfies $(p_1)$, $(p_3)$ and
$(p_5)$: There is an $r > 0$ such that $p(r) = 0$ and $p(\xi) > 0$ in $(0,r)$.
Let $\bar{p}(\xi) = p(\xi)$ if $\xi \in [0,r]$ and $\bar{p}(\xi) = 0$ otherwise. Then $\bar{p}$ satisfies $(p_1)$ - $(p_3)$. Moreover, if u is a solution of

$$(18)\qquad \begin{aligned} -\Delta u &= \lambda \bar{p}(u), \quad x \in \Omega \\ u &= 0, \qquad x \in \partial\Omega \end{aligned}$$

then $0 \leq u(x) \leq r$. Indeed, if $\mathfrak{D} = \{x \in \Omega \mid u(x) > r\}$, then

$$(19)\qquad \begin{aligned} -\Delta u &= 0, \quad x \in \mathfrak{D} \\ u &= r, \quad x \in \partial\mathfrak{D} \end{aligned}$$

so the maximum principle implies $u \equiv r$ in $\mathfrak{D}$ and, therefore, $\mathfrak{D} = \emptyset$. Similarly $\{x \in \Omega \mid u(x) < 0\} = \emptyset$. Thus, any solution of (18) is also a solution of (17). Let $\bar{P}$ be the primitive of $\bar{p}$. Now, in order to find solutions of (16), it suffices to find critical points of

$$J_\lambda(u) = \int_\Omega \left[\frac{1}{2} |\nabla u|^2 - \lambda \bar{P}(u)\right] dx$$

Letting $E = W_0^{1,2}(\Omega)$ as usual, $J_\lambda \in C^1(E,\mathbb{R})$ since $\bar{p}$ satisfies $(p_1)$ - $(p_2)$. Moreover, $J(0) = 0$ and as earlier $(p_3)$ implies $J_\lambda$ satisfies $(I_1)$. The form of $J_\lambda$ and the linear growth of $\bar{P}$ easily permit the verification of (PS). Hence, if $(I_2)$ holds for $J_\lambda$, the functional has a positive critical value. However, without further restrictions on $\lambda$, $(I_2)$ may not hold; in fact, it will not for small $\lambda$. Let $\varphi \in C^1(\bar{\Omega},\mathbb{R})$ such that $0 \leq \varphi \leq r$ and $\varphi \not\equiv 0$. Then

$\int_\Omega \bar{P}(\varphi)\,dx > 0$ and for $\lambda$ sufficiently large, say $\lambda > \lambda_0$, $J_\lambda(\varphi) < 0$. Thus, for such $\lambda$, choosing $e = \varphi$, $J_\lambda$ satisfies $(I_2)$ and (17) has a positive solution $\bar{u}(\lambda)$.

A second positive solution can now be produced as follows. Set

$$b_\lambda = \inf_{u \in E} J_\lambda(u).$$

For $\lambda > \lambda_0$, $b_\lambda < 0$ and it is easy to show that $b_\lambda$ is a critical value of $J_\lambda$. If $\underline{u}(\lambda)$ is a corresponding critical point, our above remarks imply $0 \leq \underline{u}(\lambda) \leq r$ and $\underline{u}(\lambda)$ satisfies (17). Since $J_\lambda(\underline{u}(\lambda)) < 0 < J_\lambda(\bar{u}(\lambda))$, $\underline{u}(\lambda)$ is nontrivial and must be distinct from $\bar{u}(\lambda)$. Thus, we have shown

<u>Theorem</u> <u>20</u>: If p satisfies $(p_1)$, $(p_3)$, and $(p_5)$, there is a $\lambda_0 > 0$ such that for $\lambda > \lambda_0$ (17) possesses at lease two distinct positive solutions.

See e.g. [3] for more details of the above proof. There are several variations and extensions of the Mountain Pass Theorem. See e.g. [1] or [12].

If I is invariant under a group $\mathfrak{G}$ of symmetries, it is often the case that it possesses multiple critical points. The simplest such example is when I is even (and $\mathfrak{G} = \{id, -id\}$). Results of this type are perhaps the most interesting applications of minimax methods to critical point theory. We will describe some results for the $\mathbb{Z}_2$ setting next. The earliest such is due to Ljusternik who proved:

<u>Theorem</u> <u>21</u>: If $f \in C^1(\mathbb{R}^n, \mathbb{R})$ and is even, then $f|_{S^{n-1}}$ possesses at least n distinct pairs of critical points.

Since f is even, critical points of $f|_{S^{n-1}}$ occur in pairs: $-u$ is a critical point whenever u is one. Clearly, f has a pair of maxima and minima on $S^{n-1}$. Theorem 21 shows that for $n > 2$, there are more critical points of $f|_{S^{n-1}}$ than these obvious ones. As in the Mountain Pass Theorem, it is possible to give a minimax characterization of the corresponding critical values. To do so and prove the theorem, one has to introduce an <u>index</u> <u>theory</u> to provide a means for measuring the size of symmetric (with respect to the origin) sets. See e.g. [2] or [3] for details.

There are many infinite dimensional versions of Theorem 21. E.g.

Theorem 22: If E is a real Hilbert space, $I \in C^1(E,\mathbb{R})$ is even, and $I|_{\partial B_1}$ satisfies (PS) and is bounded from below, then $I|_{\partial B_1}$ possesses infinitely many distinct pairs of critical points.

This result can be applied to elliptic problems of the form

$$(23) \qquad \begin{aligned} -\Delta u &= \lambda p(x,u), \quad x \in \Omega \\ u &= 0, \qquad x \in \partial\Omega \end{aligned}$$

The parameter $\lambda$ in (23) can be interpreted as a Lagrange multiplier which appears due to the constraint that $u \in \partial B_1$. In (23) we require that $p(x,\xi)\xi > 0$ if $\xi \neq 0$ and p satisfies $(p_1 - (p_2)$. E.g. if $p(x,\xi) = \xi + |\xi|^{s-1}\xi$, then (23) has a sequence of solutions $(\lambda_m,u_m)$ such that $u_m \in \partial B_1 \subset E = W_0^{1,2}(\Omega)$.

Next we discuss two situations where an unconstrained functional possesses multiple critical points.

Theorem 24 (Clark): Let E be a real Banach space and $I \in C^1(E,\mathbb{R})$ with I even and satisfying (PS). Suppose $I(0) = 0$, I is bounded from below and there is a set $K \subset E$ with K homeomorphic to $S^{m-1}$ by an odd map and $I|_K < 0$. Then I possesses at least m distinct pairs of critical points with negative critical values.

The proof of Clark's Theorem follows almost verbatum that of Theorem 21 relying again on a minimax characterization of critical values. See e.g. [11] or [3] for details. As a simple application, consider (23) with $p(x,\xi) = \xi - \xi^3$. Associated with this equation is the linear eigenvalue problem

$$(25) \qquad \begin{aligned} -\Delta v &= \mu v, \quad x \in \Omega \\ v &= 0, \quad x \in \partial\Omega. \end{aligned}$$

As is well known, (25) possesses a sequence of eigenvalues $(\lambda_k)$ with $0 < \lambda_1 < \lambda_2 \leq \lambda_3 \leq \ldots$ and $\lambda_k \to \infty$ as $k \to \infty$. We will show how Clark's Theorem gives at least m distinct pairs of nontrivial solutions of (23) provided that $\lambda > \lambda_m$.

Let $\bar{p}(\xi) = \xi - \xi^3$ for $0 \leq \xi \leq 1$; $= 0$ for $\xi \geq 1$; and $\bar{p}(\xi) = -\bar{p}(-\xi)$ if $\xi \leq 0$. Consider equation (18). As in an earlier argument any solution of (18) has $|u(x)| \leq 1$ and, hence, satisfies (23). To find solutions of (18) it suffices to find critical points of $J_\lambda$ defined earlier corresponding to (18). Note that $J_\lambda$ is even and by an earlier remark $J_\lambda \in C^1(E,\mathbb{R})$ with $E = W_0^{1,2}(\Omega)$, $J_\lambda(0) = 0$, and $J_\lambda$ satisfies (PS). The fact that $|\bar{P}(\xi)| \leq |\xi|$ easily implies that $J_\lambda$ is bounded from below. Thus, our existence assertion follows immediately from Clark's Theorem if we can find $K \subset E$ with K homeomorphic to $S^{m-1}$ by an odd map and $J|_K < 0$.

Let $v_1,\ldots,v_m$ be eigenfunctions of (25) corresponding to $\lambda_1,\ldots,\lambda_m$, respectively, and normalized so that $\|v_j\| = 1$, $j = 1,\ldots,m$. Let

$$K = \left\{ u = \sum_{i=1}^{m} a_i v_i \;\middle|\; \sum_{1}^{m} a_i^2 = \rho^2 \right\}$$

where $\rho$ is chosen so small that $u \in K$ implies $\bar{p}(u) = p(u)$. Clearly, K is homeomorphic to $S^{m-1}$ by an odd map. Moreover, since the functions $(v_j)$ are orthogonal in E and in $L^2$, and $\lambda > \lambda_m$, for $u \in K$

$$J_\lambda(u) = \frac{1}{2} \sum_{i=1}^{m} a_i^2 \left[ 1 - \lambda \int_\Omega v_i^2 dx \right] + \frac{\lambda}{4} \int_\Omega u^4\, dx$$

$$= \frac{1}{2} \sum_{i=1}^{m} a_i^2 \left[ 1 - \frac{\lambda}{\lambda_i} \right] + o(\rho^2) < 0$$

for $\rho$ sufficiently small. Thus, for such $\rho$, $J_\lambda < 0$ on K and our sketch is complete.

<u>Remark 26</u>: It is easy to see that the above proof works equally well if $\xi - \xi^3$ is replaced by $p(\xi)$ if p is odd, satisfies $(p_1)$ and $(p_5)$, and $p(\xi) - \xi$ satisfies $(p_3)$.

Our final result is a symmetric version of the Mountain Pass Theorem.

Theorem 27: Let E and I be as in Theorem 7 with I even and $(I_2)$ replaced by

$(I'_2)$ For all finite dimensional subspaces $\tilde{E} \subset E$, there exists $R(\tilde{E})$ such that $I \leq 0$ on $\tilde{E} \setminus B_{R(\tilde{E})}$.

Then I possesses an unbounded sequence of critical values.

Remark 28: Once again there is a minimax characterization of the critical values which is crucial to the proof.

As an application of Theorem 27, we have

Theorem 29: If p satisfies $(p_1)$ - $(p_4)$ and $p(x,\xi)$ is odd in $\xi$, then equation (2) possesses an unbounded (in $W_0^{1,2}(\Omega)$ or $L^\infty(\Omega)$) sequence of solutions.

Proof: I as defined in (4) satisfies the hypothesis of Theorem 7 and is even since $P(x,\xi)$ is even in $\xi$. Moreover, an argument given earlier--see (14)--shows I satisfies $(I'_2)$. Thus, by Theorem 27, I possesses an unbounded sequence of critical values

$$c_k = I(u_k) = \int_\Omega \left[\frac{1}{2} |\nabla u_k|^2 - P(x,u_k)\right] dx.$$

By (3) with $\varphi = u_k$,

$$(30) \quad \int_\Omega |\nabla u_k|^2 \, dx = \int_\Omega p(x,u_k)u_k \, dx.$$

Therefore,

$$(30) \quad c_k = \int_\Omega \left[\frac{1}{2} p(x,u_k)u_k - P(x,u_k)\right] dx.$$

If $(u_k)$ were bounded in E, (30)-(31) and $(p_4)$ show $(c_k)$ would be bounded in $\mathbb{R}$, contrary to Theorem 27. Thus, $(u_k)$ is unbounded in E and by similar reasoning in $L^\infty(\Omega)$.

Remark 32: A generalized version of Theorem 27 replaces $(I_1)$ by a weaker hypothesis which allows us to eliminate hypothesis $(p_3)$ in the above result. See [3].

References


[1] Nirenberg, L., Variational and topological methods in nonlinear problems, Bull. A.M.S., 4, (1981), 267-302.

[2] Rabinowitz, P. H., Some aspects of critical point theory, to appear Proc. 3rd International Sympos. Diff. Eq. and Diff. Geom., Changchun, China - 1982. (Also appears as University of Wisconsin Math. Research Center Tech. Sum. Rep. No. 2465.)

[3] Rabinowitz, P. H., Variational methods for nonlinear eigenvalue problems, eigenvalues of nonlinear problems (G. Prodi, editor). C.I.M.E., Edizioni Cremonese, Rome, (1974), 141-195.

[4] Berger, M. S., Nonlinearity and functional analysis, Academic Press, New York, 1978.

[5] Palais, R. S., Critical point theory and the minimax principle, Proc. Sym. Pure Math., 15, Amer. Math. Soc., Providence, R.I., (1970), 185-212.

[6] Schwartz, J. T., Nonlinear functional analysis, lecture notes, Courant Inst. of Math. Sc., New York University, 1965.

[7] Krasnoselski, M. A., Topological methods in the theory of nonlinear integral equations, Macmillan, New York, 1964.

[8] Vainberg, M. M., Variational methods for the study of nonlinear operators, Holden-Day, San Francisco, 1964.

[9] Ljusternik, L. A., and Schnirelmann, L. G., Topological methods in the calculus of variations, Hermann, Paris, 1934.

[10] Ambrosetti, A., and Rabinowitz, P. H., Dual variational methods in critical point theory and applications, J. Functional Analysis, 14, (1973), 349-381.

[11] Clark, D. C., A variant of the Ljusternik-Schnirelmann theory, Indiana Univ. Math. J., 22, (1972), 65-74.

[12] Benci, V., and Rabinowitz, P. H., Critical point theorems for indefinite functionals, Inv. Math., 52, (1979), 241-273

# ANALYTIC ASPECTS OF THE HARMONIC MAP PROBLEM

By

Richard M. Schoen
Mathematics Department
University of California
Berkeley, CA 94720

A fundamental nonlinear object in differential geometry is a map between manifolds. If the manifolds have Riemannian metrics, then it is natural to choose representaives for maps which respect the metric structures of the manifolds. Experience suggests that one should choose maps which are minima or critical points of variational integrals. Of the integrals which have been proposed, the energy has attracted most interest among analysts, geometers, and mathematical physicists. Its critical points, the harmonic maps, are of some geometric interest. They have also proved to be useful in applications to differential geometry. Particularly one should mention the important role they play in the classical minimal surface theory. Secondly, the applications to Kähler geometry given in [S], [SiY] illustrate the usefulness of harmonic maps as analytic tools in geometry. It seems to the author that there is good reason to be optimistic about the role which the techniques and results related to this problem can play in future developments in geometry.

This paper is both a survey and a research paper. It is a survey in that many of the results which are discussed are quite old and well known. We have not attempted to write a complete survey of the subject, but instead have chosen topics which we feel can be unified or simplified. In particular, we give more or less complete proofs of the results which we discuss. In Section 1 we formulate the harmonic map problem variationally, and discuss weakly harmonic maps and stationary points of the energy integral. In Section 2 we derive various a priori estimates on smooth harmonic maps. We prove a new estimate for maps with small energy into arbitrary manifolds (Theorem 2.2). This is a special case of a joint work with K. Uhlenbeck which is in preparation. We also derive a priori estimates in the interior and at the boundary for maps into manifolds of nonpositive curvature or target manifolds which support a strictly convex function. Many of these results are known (see references in Section 2), but our proofs are substantially new. We also give a new proof of the uniqueness theorem [Hr] based on the second variation of energy along suitably chosen homotopies. The same proof shows the homotopy minimizing property of harmonic maps into nonpositively curved manifolds. We then use these results in combination with the a priori estimates to give proofs of the theorems of [ES], [Hm] asserting existence of harmonic maps in a given homotopy class with prescribed Dirichlet boundary condition. Our proof again exploits the second variation of energy in a way which seems to be new.

In Section 3 we discuss the existence results which are known for minima, and various regularity results. We present a proof that a Hölder continuous harmonic map into any manifold is smooth. This result is known but difficult to find in the literature. Our proof is very simple and is based on approximation by linear harmonic functions. We also prove the regularity of stationary harmonic maps from a surface into any manifold. This theorem has not been previously published although we observed it quite some time ago. Finally we discuss (without proof) the regularity theory in the minimizing case. In Section 4 we discuss some generalizations and relations to PDE theory for nonlinear elliptic systems.

## 1. <u>Formulation of the Basic Problems</u>

Given Riemannian $(M^n,g)$, $(N^k,h)$ and a $C^1$ map $u:M\to N$, the energy density of u is given by

$$e(u) = \mathrm{Tr}_g(u^*h) = \sum_{\alpha,\beta,i,j} g^{\alpha\beta}(x)h_{ij}(u(x)) \frac{\partial u^i}{\partial x^\alpha}\frac{\partial u^j}{\partial x^\beta}$$

where $x^\alpha$, $u^i$ $1\leqslant\alpha\leqslant n$, $1\leqslant i\leqslant k$ are local coordinates. The energy functional $E(u)$ is then

$$E(u) = \int_M e(u)\, dv$$

where $dv = (\det g)^{1/2}\, dx$ is the volume element of M. From the definition of E it is clear that the natural class of maps u for which E is finite are those which are bounded and have first derivatives in $L^2$. Call this class of maps $\mathfrak{D}$, that is

$$\mathfrak{D} = L^{\infty}(M,N) \cap L_1^2(M,N).$$

The first observation one makes is that there is no obvious definition of these spaces because the expression for e(u) depends on the fact that the image of an open set in M is contained in a coordinate chart of N. Since we don't expect maps in $\mathfrak{D}$ to be a priori continuous we cannot guarantee this. The simplest way to get around this difficulty is to embed N as a submanifold of a Euclidean space $\mathbb{R}^K$. For convenience we assume that N is isometrically embedded, although one can get by with a smooth embedding. Now we can make sense of $L^{\infty}(M,N)$ as those members of $L^{\infty}(M,\mathbb{R}^K)$ which have image lying almost everywhere in N and similarly $L_1^2(M,N)$ is the subset of $L_1^2(M,\mathbb{R}^K)$ consisting of maps having image almost everywhere in N. Notice that N is defined by a system of equations of the form

$$\Phi(u) = (\Phi_1(u), \ldots, \Phi_{K-k}(u)) = 0$$

where the Jacobean matrix of $\Phi$ has rank K-k at each point u with $\Phi(u) = 0$. Thus the space $\mathfrak{D}$ is given by

$$\mathfrak{D} = L^{\infty}(M,\mathbb{R}^K) \cap L_1^2(M,\mathbb{R}^K) \cap \{u : \Phi(u(x)) = 0 \text{ a.e.} x\}.$$

In this formulation the energy functional becomes the ordinary Dirichlet integral

$$E(u) = \sum_{i=1}^{K} \int_M |\nabla u^i|^2 \, dv$$

and the harmonic map problem is to find critical points of E which respect the constraint $\Phi(u) = 0$. The Euler-Lagrange equation in coordinate form is

$$(1.1) \quad \Delta_M u^i + g^{\alpha\beta}\Gamma^i_{jl}(u(x)) \frac{\partial u^j}{\partial x^\alpha} \frac{\partial u^l}{\partial x^\beta} = 0, \; i = 1, \ldots, k$$

where $\Gamma^i_{jl}$ are the Christoffel symbols for the Levi-Civita connection on N; i.e. they are computed as nonlinear combinations of $h_{ij}$ and its first derivatives on N. The Euler-Lagrange equation for the constrained problem is clearly $(\Delta u)^T = 0$ where $(\;)^T$ means projection of a vector into the tangent space of N at u(x). Since the normal component of $\Delta u$ can be represented in terms of the second fundamental form of N, the equation becomes

$$(1.2) \quad \Delta_M u^i = g^{\alpha\beta} A^i_{u(x)} \left[\frac{\partial u}{\partial x^\alpha}, \frac{\partial u}{\partial x^\beta}\right], \; i = 1, \ldots, K$$

where $A_u(X,Y) \in (T_u N)^\perp$ is the second fundamental form of N given by $A(X,Y) = (D_X Y)^\perp$ where D denotes the directional derivative in $\mathbb{R}^K$ and X,Y are extended arbitrarily as tangent vector fields to N in a neighborhood of $u \in N$. The advantage of (1.2) over (1.1) is that (1.2) makes sense in weak form for a map in $\mathfrak{D}$.

**Definition.** A map $u \in \mathfrak{D}$ is a <u>(weakly) harmonic map</u> if u satisfies the weak form of (1.2); that is for any $\eta \in C_0^\infty(M, \mathbb{R}^K)$ we have

$$(1.3) \quad \int_M \sum_i \nabla\eta^i \cdot \nabla u^i + g^{\alpha\beta}\eta^i(x) \, A^i_{u(x)} \left[\frac{\partial u}{\partial x^\alpha}, \frac{\partial u}{\partial x^\beta}\right] dv = 0.$$

One can also ask that u be a critical point of E in $\mathfrak{D}$. Observe, however, that since $\mathfrak{D}$ is only defined as a subset of the Hilbert space $L_1^2(M, \mathbb{R}^K)$, there is no clear meaning of critical point. We formulate a definition based on utility. There are two basic classes of variations which one would like to perform on a map u.

The first is defined by a function $\eta \in C_0^\infty(M,\mathbb{R}^K)$. Given such an $\eta$ we can consider the map $u + t\eta$. Unless we have amazingly good luck the map $u + t\eta$ will not be in $\mathfrak{D}$ because it will not satisfy the constraint $\Phi(u) = 0$. If N is at least a $C^1$ submanifold we can correct for this deficiency by projecting $u + t\eta$ onto N. Precisely let $\Pi: O \to N$ be the nearest point projection defined on a small neighborhood O of N and observe that since u has bounded image we will have $u(x) + t\eta(x) \in O$ for t sufficiently, a.e. $x \in M$. Therefore we can get $u_t = \Pi \cdot (u + t\eta)$ and it is not difficult to see that the curve $t \to u_t$ is a $C^1$ curve in $L_1^2(M,\mathbb{R}^K)$ and hence we can require

$$\frac{d}{dt} E(u_t)\Big|_{t=0} = 0 \text{ for every } \eta \in C_0^\infty(M,\mathbb{R}^K).$$

It can be seen without much difficulty (see [SU1]) that u is harmonic if and only if u is critical for these variations. There is a second type of variation which one can consider. This is a variation of the parameter domain M; that is, given a family $F_t: M \to M$ of diffeomorphisms which are all equal to the identity outside a compact set of M and with $F_0 = \text{id}$, we can set $u_t = u \circ F_t$ and require that

$$\frac{d}{dt} E(u_t)\Big|_{t=0} = 0 \text{ for every family } F_t.$$

It can be seen that the function $t \to E(u_t)$ is a $C^1$ function of t so that the derivative exists. It does not seem to be true that the curve $t \to u_t$ is a differentiable curve in $L_1^2(M,\mathbb{R}^K)$.

**Definition.** A map $u \in \mathfrak{D}$ is a stationary point of E if u is critical with respect to both of the types of variations described above.

As we have noted, a stationary point is harmonic but the converse is not known. It is clear that a $C^2$ harmonic map is a stationary point; in fact we'll see below that a $C^0$ harmonic map is a stationary point. The following lemma is essentially a classical result.

**Lemma 1.1.** *Suppose* $n = \dim M = 2$, *and* $u \in \mathfrak{D}$ *is a stationary point for* E. *The Hopf differential*

$$\Phi(z) = \left[\left(\|u_*\left(\frac{\partial}{\partial x}\right)\|^2 - \|u_*\left(\frac{\partial}{\partial y}\right)\|^2\right) - 2i < u_*\left(\frac{\partial}{\partial x}\right), u_*\left(\frac{\partial}{\partial y}\right) >\right] dz^2$$

*where* $z = x + iy$ *is a complex coordinate on* M, *is holomorphic and hence smooth.*

**Proof.** If the map u is $C^2$, this result was derived in [CG]. Since the result is local, we can assume that the domain of u is the unit disk in $\mathbb{C}$. For any $C^\infty$ function $\eta(x,y)$ with compact support in D, and for t small, set $F_t(x,y) = (x + t\eta(x,y),y)$, and $u_t = u \circ F_t$. By the chain rule for weak derivatives we have

$$\frac{\partial u_t}{\partial x} = \frac{\partial u}{\partial x}(F_t(x,y))\left[1 + t\frac{\partial \eta}{\partial x}\right]$$

$$\frac{\partial u_t}{\partial y} = \frac{\partial u}{\partial x}(F_t(x,y))\left[t\frac{\partial \eta}{\partial y}\right] + \frac{\partial u}{\partial y}(F_t(x,y)).$$

Making the change of variable $(\xi,\tau) = F_t(x,y)$ we then have

$$E(u_t) = \int_D \left[\left|\frac{\partial u}{\partial \xi}\right|^2 \left[1 + t\frac{\partial \eta}{\partial x}\right]^2 + \left|\frac{\partial u}{\partial \xi} \cdot t\frac{\partial \eta}{\partial y} + \frac{\partial u}{\partial \tau}\right|^2\right] \frac{d\xi d\tau}{(1+t\eta_x)}$$

Differentiating with respect to t and setting t = 0 we get

$$0 = \int_D \left[\left(\left|\frac{\partial u}{\partial x}\right|^2 - \left|\frac{\partial u}{\partial y}\right|^2\right)\frac{\partial \eta}{\partial x} + 2\left<\frac{\partial u}{\partial x}, \frac{\partial u}{\partial y}\right>\frac{\partial \eta}{\partial y}\right] dxdy$$

for every smooth $\eta$ of compact support in D. By a similar argument making use of the diffeomorphism $G_t(x,y) = (x,y + t\eta(x,y))$ we have

$$\int_D \left[\left(\left|\frac{\partial u}{\partial x}\right|^2 - \left|\frac{\partial u}{\partial y}\right|^2\right)\frac{\partial \eta}{\partial y} - 2\left<\frac{\partial u}{\partial x}, \frac{\partial u}{\partial y}\right>\frac{\partial \eta}{\partial x}\right] dxdy = 0.$$

for all $\eta$. These equations are the weak form of the Cauchy-Riemann equations for the $L^1$ function

$$\varphi(z) = \left(\left|\frac{\partial u}{\partial x}\right|^2 - \left|\frac{\partial u}{\partial y}\right|^2\right) - 2i\left<\frac{\partial u}{\partial x}, \frac{\partial u}{\partial y}\right>.$$

By Weyl's lemma, $\varphi(z)$ is a smooth holomorphic function of z. This completes the proof Lemma 1.1.

<u>Remark</u> <u>1.2</u>. The smoothness of the Hopf differential does not seem to follow from the condition that u be a harmonic map. One needs to use variations of the parameter domain which cannot be justified from (1.3) alone.

In dimension greater than two, stationary maps enjoy the following monotonicity inequality. Suppose u is a stationary map from a domain $\Omega$ in $\mathbb{R}^n$ into a manifold N. Then we have for $0<\sigma<\rho<\text{dist}(x_0,\partial\Omega)$

$$(1.4)\quad \rho^{2-n}\int_{B_\rho(x_0)} e(u)dx - \sigma^{2-n}\int_{B_\sigma(x_0)} e(u)dx$$

$$= \int_{B_\rho(x_0)-B_\sigma(x_0)} |x-x_0|^{2-n}\left|\frac{\partial u}{\partial r}\right|^2 dx$$

where $r = |x-x_0|$. For stationary maps (1.4) was derived in [P]. For maps of least energy it was used in the regularity theory in [SU1].

To close this introductory section we record the formula for the second derivative of energy for a $C^2$ variation through smooth maps. We will exploit this in the next section. Consider a $C^2$ map $F:M\times[0,1]\to N$, and set $u_t(x) = F(x,t)$ for $x\in M$. The following formulas are derived for example in [EL],

$$(1.5)\quad \frac{d}{dt}E(u_t) = 2\int_M \sum_\alpha \langle\nabla'_{e_\alpha}V,F_*(e_\alpha)\rangle\, dv_M$$

$$(1.6)\quad \frac{d^2}{dt^2}E(u_t) = 2\int_M \Big[\sum_\alpha \|\nabla'_{e_\alpha}V\|^2 - \langle R^N(V,F_*e_\alpha)V,F_*e_\alpha\rangle$$

$$+ \langle\nabla'_{e_\alpha}\nabla'_{\frac{\partial}{\partial t}}V,F_*e_\alpha\rangle\Big]\, dv$$

Here we have set $V = F_*\left[\frac{\partial}{\partial t}\right]$ and have denoted by $\nabla'$ the pullback connection from TN. The next lemma gives conditions under which the third term on the right can be ignored.

**Lemma 1.3.** *Suppose either of the following holds:*

(i) *The map* $u = u_0$ *is harmonic, and* $V(x,0)$ *has compact support in the interior of* M, *or*

(ii) *For each* $x \in M$, *the curve* $\gamma_x(t) = F(x,t)$ *is a constant speed geodesic in* N.

*Then we have*

$$\frac{d^2}{dt^2} E(u_t)\Big|_{t=0} = 2\int_M \left[ \|\nabla' V\|^2 - \sum_\alpha \langle R^N(V,u_* e_\alpha)V, u_* e_\alpha\rangle \right] dv_M$$

**Proof.** If the map u is harmonic, then the third term on the right of (1.6) vanishes because it represents the first derivative of energy with respect to an initial variation given by $\nabla'_{\frac{\partial}{\partial t}} V$. On the other hand, if the curves $\gamma_x(t)$ are constant speed geodesics, then we have $\nabla'_{\frac{\partial}{\partial t}} V \equiv 0$. This proves Lemma 1.3.

## 2. A Priori Estimates and Special Target Manifolds

In this section we consider estimates which can be derived for smooth harmonic maps. These estimates can be exploited to prove existence theorems for maps into manifolds N of nonpositive curvature. The Bochner formula for harmonic maps (see [EL] for a derivation) says

$$(2.1)\quad \frac{1}{2}\Delta e(u) = \|\nabla' du\|^2 - \sum_{\alpha,\beta} \langle R^N(u_* e_\alpha, u_* e_\beta)u_* e_\alpha, u_* e_\beta\rangle$$
$$+ \sum_i \mathrm{Ric}_M(u^*\theta_i, u^*\theta_i)$$

where $e_1, \ldots, e_n$ is an orthonormal basis for TM and $\theta_1, \ldots, \theta_k$ is orthonormal for $T^*N$. We see from (2.1) that if M is compact (with possibly empty boundary) and N has nonpositive sectional curvature, then

$$(2.2) \qquad \Delta e(u) \geq -Ce(u)$$

with C depending only on M. A standard result in PDE (see [M3,5.3.1]) then implies the estimate

$$(2.3) \qquad \sup_{\Omega} e(u) \leq C \int_M e(u)\, dv$$

for any open $\Omega$ compactly contained in M, where C depends only on M, n, and dist($\Omega,\partial M$). The estimate (2.3) was first proved for the heat equation associated to the harmonic map problem in [ES]. The estimate (2.3) implies the following compactness result.

**Proposition 2.1.** *If M is compact with (possibly empty) boundary, and N is compact with nonpositive sectional curvatures, then the family of maps, for* $\Lambda>0$

$$\mathcal{F}_\Lambda = \{u \in C^\infty(M,N): u \text{ is harmonic, } E(u) \leq \Lambda\}$$

*is compact in the topology of uniform* $C^k$ *convergence on compact subsets of M for any integer* $k \geq 0$.

The proof of Proposition 2.1 involves deriving estimates on all derivatives of maps $u \in \mathcal{F}_\Lambda$ locally on M, and applying Ascoli's theorem on equicontinuity. It is a standard application of linear elliptic theory to show that higher derivative estimates follow from (2.3). The inequality (2.2) relies heavily on the hypothesis that N have nonpositive sectional curvature. Conformal maps of $S^2$ give examples which show that (2.3) is false for maps into arbitrary compact manifolds. The following theorem which is due to K. Uhlenbeck and the author shows that (2.3) does in fact hold for harmonic maps with small total energy.

**Theorem** 2.2. *Suppose* $u \in C^2(B_r, N)$ *is harmonic with respect to a metric* $g$ *on* $B_r = \{x \in \mathbb{R}^n : |x| < r\}$ *which satisfies*

$$\Lambda^{-1}(\delta_{ij}) \leqslant (g_{ij}) \leqslant \Lambda(\delta_{ij}), \quad |\partial_k g_{ij}| \leqslant \Lambda r^{-1}.$$

*There exists* $\epsilon > 0$ *depending only on* $\Lambda$, $n$, $N$ *such that if*

$$r^{2-n} \int_{B_r} e(u) \leqslant \epsilon,$$

*then* $u$ *satisfies the inequality*

$$\sup_{B_{r/2}} e(u) \leqslant C\, r^{-n} \int_{B_r} e(u).$$

*where* $C$ *depends only on* $\Lambda$, $n$, $N$.

**Remark** 2.3. The proof we will give actually only requires that the $g_{ij}$ be Hölder continuous on $B_r$.

**Proof**. The monotonicity inequality (1.4) implies

$$(2.4) \qquad \sigma^{2-n} \int_{B_\sigma(x)} e(u) \leqslant C\, \rho^{2-n} \int_{B_\rho(x)} e(u)$$

provided $x \in B_r$, and $0 < \sigma < \rho < r - |x|$. If we set $r_1 = \frac{3}{4} r$, then inequality (2.4) implies

$$(2.5) \qquad \sigma^{2-n} \int_{B_\sigma(x)} e(u) \leqslant \left(\frac{4}{3}\right)^{n-2} C\, r^{2-n} \int_{B_r} e(u)$$

for any $x \in B_{r_1}$, $0 < \sigma \leqslant r_1 - |x|$. In (2.4) and (2.5) the constant $C$ depends only on $n, \Lambda$.

We will exploit (2.5) for $\sigma$ small. First observe that there exists $\sigma_0 \in (0,r_1)$ such that

$$(r_1 - \sigma_0)^2 \sup_{B_{\sigma_0}} e(u) = \max_{0<\sigma\leqslant r_1} (r_1 - \sigma)^2 \sup_{B_\sigma} e(u).$$

Moreover, there exists a point $x_0 \in \bar{B}_{\sigma_0}$ such that

$$e_0 = e(u)(x_0) = \sup_{B_{\sigma_0}} e(u).$$

Set $\rho_0 = \frac{1}{2}(r_1 - \sigma_0)$, and observe that from the choice of $\sigma_0, x_0$

(2.6) $$\sup_{B_{\rho_0}(x_0)} e(u) \leqslant \sup_{B_{\sigma_0+\rho_0}} e(u) \leqslant 4e_0$$

Now define a map $v \in C^2(B_{r_0},N)$, $r_0 = (e_0)^{1/2} \rho_0$, by

$$v(y) = u\left(\frac{y-x_0}{(e_0)^{1/2}}\right).$$

Thus v is a rescaled version of u chosen so that $e(v)(0) = 1$. From (2.6) we get

$$\sup_{B_{r_0}} e(v) \leqslant 4, \quad e(v)(0) = 1.$$

Therefore we see from (2.1) that $\Delta e(v) \geqslant -Ce(v)$ on $B_{r_0}$. Now if $r_0 \geqslant 1$, then we can apply the mean value inequality to get

$$1 = e(v)(0) \leqslant C \int_{B_1} e(v).$$

However, we have from (2.5)

$$\int_{B_1} e(v) = \left[(e_0)^{1/2}\right]^{n-2} \int_{B_{e_0^{-1/2}}(x_0)} e(u)$$

$$\leqslant C\, r^{2-n} \int_{B_r} e(u) \leqslant C\varepsilon.$$

Thus if $\varepsilon$ is chosen small enough, these two inequalities are in contradiction. Therefore we may assume $r_0 \leqslant 1$. It then follows from [M3,5.3.1] that the mean value inequality holds for $e(v)$ on $B_{r_0}$, and we have

$$1 = e(v)(0) \leqslant C\, r_0^{-n} \int_{B_{r_0}} e(v) = C\, r_0^{-2}\, \rho_0^{2-n} \int_{B_{\rho_0}(x_0)} e(u).$$

Combining this with (2.5) we get

$$\rho_0^2\, e_0 = r_0^2 \leqslant C\, r^{2-n} \int_{B_r} e(u).$$

From the choice of $\sigma_0$ this implies

$$\max_{0<\sigma\leqslant r_1} (r_1 - \sigma)^2 \sup_{B_\sigma} e(u) \leqslant 4\, C\, r^{2-n} \int_{B_r} e(u).$$

The conclusion of Theorem 2.2 now follows by taking $\sigma = \frac{1}{2} r$ and dividing by $r^2$.

__Corollary 2.3__. *Let* M *be compact with possibly empty boundary, and let* N *be compact. For* $\Lambda>0$, *set*

$$\mathcal{F}_\Lambda = \{u \in C^\infty(M,N) : u \text{ is harmonic, } E(u) \leqslant \Lambda\}.$$

*Any map u in the weak $L_1^2$ closure of $\mathcal{F}_\Lambda$ is smooth and harmonic outside a closed singular set of locally finite Hausdorf (n-2)-dimensional measure.*

**Proof.** Let $\{u_i\} \subset \mathcal{F}_\Lambda$ be a sequence converging weakly to u.

Let $\overset{\circ}{M}$ denote the interior of M, and define a subset $\mathcal{S} \subset \overset{\circ}{M}$ by

$$\mathcal{S} = \bigcap_{r>0} \{x \in \overset{\circ}{M} : \underline{\lim}_{i\to\infty} r^{2-n} \int_{B_r(x)} e(u_i) \geq \varepsilon_0\}.$$

Here $\varepsilon_0$ will denote a fixed positive number determined by Theorem 2.2. It follows from the monotonicity of the quantity $r^{2-n} \int_{B_r} e(u_i)$ that $\mathcal{S}$ is a relatively closed subset of $\overset{\circ}{M}$, for suppose $x_j \in \mathcal{S}$ and $\lim x_j = x \in \overset{\circ}{M}$. For any $r > 0$ and all j, it follows from (1.4) that

$$\underline{\lim}_{i\to\infty} r^{2-n} \int_{B_r(x_j)} e(u_i) \geq \varepsilon_0.$$

This immediately implies $\underline{\lim}_{i\to\infty} r_1^{2-n} \int_{B_{r_1}(x)} e(u_i) \geq \varepsilon_0$ for any $r_1 > r$. Since $r > 0$ was arbitrary we have $x \in \mathcal{S}$.

For any $\delta > 0$, and any $\Omega \subset\subset \overset{\circ}{M}$ we can cover $\mathcal{S} \cap \bar{\Omega}$ by a finite collection of balls $\{B_{r_j}(x_j)\}$ so that the collection $\{B_{\frac{1}{2}r_j}(x_j)\}$ is disjoint, and so that $x_j \in \mathcal{S}$. For i sufficiently large we then have

$$\left(\frac{1}{2}r_j\right)^{2-n} \int_{B_{\frac{1}{2}r_j}(x_j)} e(u_i) \geq \varepsilon_0 \text{ for all j.}$$

Summing on j we then get

$$\sum_j r_j^{n-2} \leqslant C\ E(u_i) \leqslant C\ \Lambda.$$

It follows that $\mathcal{H}^{n-2}(\mathcal{S}\cap\bar{\Omega}) \leqslant C\ \Lambda$.

We show that a subsequence of $\{u_i\}$ converges uniformly in $C^k$ norm to u on compact subsets of $\overset{\circ}{M}-\mathcal{S}$. Let $x\in\overset{\circ}{M}-\mathcal{S}$, and observe that for some r we have

$$r^{2-n} \int_{B_r(x)} e(u_i) < \varepsilon_0$$

for infinitely many i. By Theorem 2.2 we have uniform $C^1$, and hence $C^k$, estimates on $u_i$ in $B_{\frac{1}{2}r}(x)$. Thus a subsequence of $\{u_i\}$ converges uniformly in $C^k$ norm in $B_{\frac{1}{2}r}(x)$. By uniqueness of weak limits, the limit is u. Therefore u is regular outside $\mathcal{S}$, and by a diagonal subsequence argument a subsequence of $\{u_i\}$ converges uniformly in $C^k$ norm to u on compact subsets of $\overset{\circ}{M}-\mathcal{S}$. This completes the proof of Corollary 2.3.

Theorem 2.2 requires the smallness of energy as a hypothesis. While this hypothesis is not generally satisfied at every point for maps into an arbitrary manifold (e.g. conformal maps of $S^2$), under certain assumptions on the target, it can be verified. In particular, if there is a smooth bounded strictly convex function g defined on N, we can verify the hypothesis. The following argument appears in [GH]. To see this, observe that $g\circ u$ is subharmonic, in fact we may assume

$$(2.7)\qquad \Delta(g\circ u) \geqslant \varepsilon_1\ e(u), \quad 0 \leqslant g\circ u \leqslant C.$$

Given $x\in\overset{\circ}{M}$, $B_r(x) \subset \overset{\circ}{M}$, we can apply Green's formula to $\varphi(x)g\circ u$ where $\varphi$ has compact support in $B_r(x)$ with $\varphi(x) = 1$

$$g \circ u(x) = \int_{B_r(x)} G(x,y)\ \Delta(\varphi g \circ u(y))\ dy.$$

Choosing $\varphi$ to be equal to 1 on $B_{\frac{1}{2}r}(x)$ with compact support, we can easily see from (2.7)

$$\int_{B_{\frac{1}{2}r}(x)} G(x,y)\ e(u)(y)\ dy \leqslant C.$$

In particular, this implies (assume $n > 2$, for one handles the case $n = 2$ by a simple modification of these arguments.)

$$\int_{B_{\frac{1}{2}r}(x)} d(x,y)^{2-n}\ e(u)(y)\ dy \leqslant C.$$

On the other hand, if we set $R(\sigma) = \sigma^{2-n} \int_{B_\sigma(x)} e(u)$, we have

$$\int_0^{\frac{1}{2}r} \frac{R(\sigma)}{\sigma}\ d\sigma = (2-n)^{-1} \int_0^{\frac{1}{2}r} \frac{d}{d\sigma}(\sigma^{2-n}) \left[ \int_{B_\sigma(x)} e(u) \right] d\sigma$$

$$\leqslant (n-2)^{-1} \int_{B_{\frac{1}{2}r}(x)} d(x,y)^{2-n}\ e(u)(y)\ dy$$

Since $R(\sigma)$ is a monotone function of $\sigma$, we have for any $\sigma_0 \in (0, \frac{1}{2}r)$

$$R(\sigma_0)\log(\sigma_0^{-1}\tfrac{1}{2}r) \leq C.$$

This shows that if $\sigma_0$ is chosen small, then $R(\sigma_0)$ is small, and Theorem 2.2 applies in $B_{\sigma_0}(x)$. We have therefore shown.

**Corollary 2.4.** *If* M *is as above, and* $u \in C^\infty(M,N)$ *is harmonic such that a bounded strictly convex function* g *exists on* u(M), *then for* $\Omega \subset\subset M$ *we have*

$$\|u\|_{C^k(\Omega)} \leq C$$

*where* C *depends only on* k, n, M, N, $\Omega$, *and* E(u). *In particular, Proposition* 2.1 *holds under this convex function hypothesis.*

**Remark 2.5.** Existence, regularity, and a priori estimates have been given in [HKW,HW] under the assumption that the image of u lies in a strictly convex ball. In fact, the results of [HW] also apply to the more difficult case when the metric of M is assumed only bounded, measurable, and (locally) uniformly Euclidean.

The interior a priori estimates given above can be derived at the boundary under the Dirichlet boundary condition. We state the boundary analogue of Theorem 2.2. Since the proof of Theorem 2.2 uses only (1.4) together with linear elliptic estimates (more precisely, estimates on harmonic maps with bounded energy density), the proof works with minor changes at the boundary. Suppose M is a compact manifold with smooth boundary $\partial M$. For a point $x \in \partial M$, we use $B_r^+(x)$ to denote

$$B_r^+(x) = \{y \in M \cup \partial M : d(y,x) < r\}.$$

We impose the Dirichlet boundary condition in $B_r^+(x)$, that is, we assume $\varphi \in C^2(B_r^+(x),N)$ satisfies

$$r|D\varphi| + r^2|DD\varphi| \leq \Lambda, \quad u = \varphi \text{ on } \partial M \cap B_r^+(x).$$

We also assume the metric g satisfies in $B_r^+(x)$

$$\Lambda^{-1}(\delta_{ij}) \leqslant (g_{ij}) \leqslant \Lambda(\delta_{ij}), \qquad r|\partial_k g_{ij}| \leqslant \Lambda$$

as in Theorem 2.2. We then have

**<u>Theorem</u> <u>2.6</u>.** *Suppose* $u \in C^2(B_r^+, N)$ *is harmonic with Dirichlet data* $\varphi$. *There exists* $\varepsilon > 0$ *depending only on* $\Lambda$, n, N *such that if*

$$r^{2-n} \int_{B_r^+(x)} e(u) \leqslant \varepsilon,$$

*then* u *satisfies the inequality*

$$\sup_{B_{r/2}^+} e(u) \leqslant C\left[ r^{-n} \int_{B_r^+} e(u) + \sup_{B_r^+} e(\varphi) \right]$$

*with* C *depending only on* $\Lambda$, n, N.

As above, the smallness assumption can be removed under appropriate hypotheses on N.

**<u>Theorem</u> <u>2.7</u>.** *Suppose either* N *has nonpositive sectional curvature or the image of* u *carries a bounded strictly convex function. Then, if* u *and* $\varphi$ *are as above, we have*

$$\sup_{B_{r/2}^+} e(u) \leqslant C\left[ r^{-n} \int_{B_r^+} e(u) + \sup_{B_r^+} e(\varphi) \right]$$

*with* C *depending only on* $\Lambda$, n, N.

**<u>Remark</u>.** The a priori boundary estimate of Theorem 2.7 under the assumption that N have nonpositive curvature can also be derived from the results of [HKW]. One can lift u, $\varphi$ to maps from $B_r^+(x)$

to the universal cover of N. The Poincare inequality can then be used to show

$$\int_{B_r^+(x)} \rho^2 \leqslant C\left[ r^2 \int_{B_r^+} e(u) + r^n \sup_{B_r^+} e(\varphi) \right]$$

where $\rho(y) = \mathrm{dist}_{\tilde{N}}(u(x),u(y))$. Since $\rho^2$ is a subharmonic function which is bounded on $\partial M \cap B_r^+$, we then get a point-wise bound on $\rho^2$ and hence the theory of [HKW] can be applied.

__Corollary 2.8.__ *Suppose* $\varphi \in C^2(M \cup \partial M, N)$, *and* u *is a* $C^2$ *harmonic map agreeing with* $\varphi$ *on* $\partial M$. *If* N *has nonpositive sectional curvature or* u(M) *supports a strictly convex function, then the following holds*

$$\sup_{M \cup \partial M} e(u) \leqslant C\left[ \int_M e(u) + \sup_{M \cup \partial M} e(\varphi) \right]$$

*with* C *depending only on* n, M, N, *and* $\|\varphi\|_{C^2}$.

We will use these estimates below to prove the existence of harmonic maps with Dirichlet boundary condition in a given homotopy class. Our proof will also use certain uniqueness results which we now describe. Suppose M is a compact manifold with boundary, and denote $\bar{M} = M \cup \partial M$. If $\varphi \in C^2(\bar{M},N)$ is a given map, we consider maps $u \in C^2(\bar{M},N)$ which agree with $\varphi$ on $\partial M$ and are homotopic to $\varphi$, denoted $u \sim \varphi$. By this we mean there is a $C^2$ map $F: M \times [0,1] \to N$ with

$F(x,t) = \varphi(x)$ for all $x \in \partial M$, $t \in [0,1]$

$F(x,0) = u(x)$ for all $x \in \bar{M}$

$F(x,1) = \varphi(x)$ for all $x \in \bar{M}$.

The following result is a consequence of the theorems of Eells-Sampson [ES], Hamilton [Hm], and Hartman [Hr]. We give a simple proof based on the idea of [ScY1].

**Theorem 2.9.** *Suppose* $u, \varphi$ *are as above with* $u \sim \varphi$ *and suppose* $u$ *is harmonic. Suppose* $N$ *is compact with nonpositive curvature. Then* $u$ *minimizes energy over all maps homotopic to* $\varphi$ *and equal to* $\varphi$ *on* $\partial M$.

**Proof.** It suffices to show that $E(u) \leq E(\varphi)$. Let $F: M \times [0,1] \to N$ be a homotopy from $u$ to $\varphi$. We replace $F$ by a geodesic homotopy as follows. For any $x \in \bar{M}$, let $\gamma_x(s)$ denote the unique geodesic arc in $N$, parameterized with constant speed (depending on $x$) for $s \in [0,1]$, and connecting $u(x)$ with $\varphi(x)$ and homotopic to the curve $t \mapsto F(x,t)$. The uniqueness of $\gamma_x$ follows because $N$ has nonpositive curvature. Define $G: M \times [0,1] \to N$ by $G(x,s) = \gamma_x(s)$, and let $u_s \in C^2(\bar{M},N)$ be given by $u_s(x) = G(x,s)$. Thus we have $u_0 = u$ and $u_1 = \varphi$. Since the $s$ parameter curves of the homotopy $G$ are constant speed geodesics, we see from Lemma 1.3

$$\frac{d^2}{ds^2} E(u_s) = 2 \int_M \left(\|\nabla' V\|^2 - \sum_\alpha \langle R^N(V, G_* e_\alpha)V, G_* e_\alpha \rangle\right) dv_M$$

where $e_1, \ldots, e_n$ is an orthonormal basis for $TM$, and $V = G_*\left[\frac{\partial}{\partial s}\right]$. In particular, since $N$ has nonpositive curvature, we see that the function $s \mapsto E(u_s)$ is a convex function on $[0,1]$. Since $u = u_0$ is harmonic, we have $\frac{d}{ds} E(u_s) = 0$ at $s = 0$. Therefore $E(u_s)$ is nondecreasing in $s$ and hence $E(u) \leq E(\varphi)$ as required.

The same argument gives a proof of the uniqueness theorem of Hartman [Hr].

**Theorem 2.10.** *Suppose* $u_0, u_1$ *are homotopic harmonic maps which agree on* $\partial M$. *If* $N$ *has nonpositive curvature and* $\partial M \neq \emptyset$, *then* $u_0 \equiv u_1$ *on* $M$. *If* $\partial M = \emptyset$, $N$ *has negative curvature, and* $u_0$ *has rank greater than one at some point of* $M$, *then* $u_0 \equiv u_1$ *on* $M$.

**Proof.** Let $u_s$, $s \in [0,1]$ be the geodesic homotopy constructed above. Since the function $s \mapsto E(u_s)$ is convex and has slope zero at 0 and 1 ($u_0, u_1$ are harmonic), we must have $\frac{d^2}{ds^2} E(u_s) \equiv 0$ for $0 \leq s \leq 1$. Therefore

$$(2.8) \quad \nabla' V \equiv 0, \quad \sum_\alpha \langle R^N(V, G_* e_\alpha)V, G_* e_\alpha \rangle \equiv 0$$

on $M \times [0,1]$. Now we have

$$e_\alpha \|V\|^2 = 2 \langle V, \nabla'_{e_\alpha} V \rangle = 0, \quad \alpha = 1, \ldots, n$$

and hence $\|V\|$ is constant in x for each s. If $\partial M \neq \emptyset$, then $V = 0$ on $\partial M \times [0,1]$ and hence $V \equiv 0$ on $M \times [0,1]$. Thus $u_0 \equiv u_1$ on M. If $\partial M = \emptyset$ and N has negative curvature, then from (2.8) we see that V is parallel to $G_* e_\alpha$ for $\alpha = 1, \ldots, n$. If V is not identically zero at $s = 0$, then $V(x,0)$ is nowhere zero since $\|V\|$ is constant in x. Then it follows that $u_0$ has rank 1 everywhere on M, a contradiction. Therefore $V(x,0) \equiv 0$, and hence $V(x,s) \equiv 0$ for all $(x,s) \in M \times [0,1]$ because the geodesics $s \mapsto G(x,s)$ are determined by their initial tangent vector. This proves Theorem 2.10.

We now treat the existence theorems for maps into manifolds of nonpositive curvature. An application of the Leray-Schauder degree theory based on the a priori estimates above can be used to prove the following.

<u>**Lemma**</u> <u>2.11</u>. *Suppose* M *is a compact simply connected manifold with boundary. Given* $\varphi \in C^3(\bar{M}, N)$, *where* N *is compact with nonpositive curvature, there is a unique harmonic map* $u \in C^2(\bar{M}, N)$ *with* $u = \varphi$ *on* $\partial M$.

<u>**Proof**</u>. We may lift $\varphi$ to a map $\tilde{\varphi}: M \to \tilde{N}$ where $\tilde{N}$ is the universal cover of N. Let $h = (h_{ij})$ be the given metric on $\tilde{N}$ written in normal coordinates u about some point. Thus we have

$$h_{ij}(0) = \delta_{ij}, \quad \partial_k h_{ij}(0) = 0.$$

For $t \in [0,1]$, define a metric ${}^t h$ on $\tilde{N}$ by ${}^t h(u) = h(tu)$. Clearly ${}^t h$ has nonpositive curvature for $t \in [0,1]$, and ${}^\circ h$ is the Euclidean metric. Since we know the linear existence theory for ${}^\circ h$, we can use the a priori estimates of Corollary 2.4 together with a continuity method (or degree theory) to prove Lemma 2.11.

We now give a proof of the theorems of Eells-Sampson [ES] and Hamilton [Hm] based on the "local" existence theorem of the previous lemma together with the a priori estimates and Theorem 2.9.

**Theorem 2.12.** *Suppose M is a compact manifold with (possibly empty) boundary, and suppose N is compact with nonpositive sectional curvature. Given* $\varphi \in C^3(\bar{M},N)$, *there is a harmonic map* $u \in C^2(\bar{M},N)$ *such that* $u = \varphi$ *on* $\partial M$, *and u is homotopic to* $\varphi$ *through maps with fixed values on* $\partial M$.

**Proof.** Let $E_0$ be the infimum of energy taken over all smooth maps homotopic to $\varphi$ with fixed boundary values. Let $\{u_i\}$ be a sequence of such maps such that

$$\lim_{i\to\infty} E(u_i) = E_0.$$

By weak compactness of bounded sets in $L_1^2(M,N)$, a subsequence of $\{u_i\}$, again denoted $\{u_i\}$, converges weakly to a limit $u_0 \in L_1^2(M,N)$. We show that $u_0$ is a smooth harmonic map. First take a point $x \in \overset{\circ}{M}$, the interior of M and choose a ball B centered at x contained in $\overset{\circ}{M}$. Apply Lemma 2.11 to construct a harmonic map $h_i \in C^2(B,N)$ with $h_i = u_i$ on $\partial B$. By the compactness result, Proposition 2.1, a subsequence of $\{h_i\}$ converges to a smooth limit h on compact subsets of B. We will prove that h = u a.e. in a neighborhood of x. To see this, let $h_s$ be the geodesic homotopy from $h_i$ to $u_i$ (see Theorem 2.9). We have for $s \in [0,1]$

$$\frac{d^2}{ds^2} E(u_s) \geq 2 \int_B \|\nabla' V\|^2$$

where $V = G_*\left(\frac{\partial}{\partial s}\right)$, $G(x,s) = h_s(x)$. Notice that by the compatibility of $\nabla'$ with the inner product we have

$$|\nabla\|V\||^2 \leq \|\nabla' V\|^2.$$

Also we have $V = 0$ on $\partial M \times [0,1]$, and $\|V\|(x,s)$ is independent of s since $\|V\|$ represents the length of the tangent vector to a constant speed geodesic. It follows from the Poincare inequality

$$\int_B \|V\|^2 \leqslant C \int_B \|\nabla' V\|^2.$$

Since $E(h_s)$ has first derivative zero at $s = 0$, it follows from the fundamental theorem of calculus that for some $s_0 \in [0,1]$ we must have at $s = s_0$.

$$\frac{d^2}{ds^2} E(h_s) \leqslant E(h_1) - E(h_0).$$

On the other hand, by Theorem 2.9 and the definition of $E_0$

$$E(h_1) - E(h_0) = \int_M e(u_i) - \int_M e(\hat{u}_i) \leqslant E(u_i) - E_0$$

where $\hat{u}_i$ is the map which equals $u_i$ on M–B and $h_i$ on B. Combining the above inequalities we have

$$\int_B \|V\|^2 \leqslant C\ (E(u_i) - E_0).$$

Now recall that $N \subset \mathbb{R}^K$ isometrically, so we have

$$|u_i - h_i| \leqslant d(u_i, h_i)$$

where $d(\cdot,\cdot)$ denotes geodesic distance on N. Also, note that $\|V\|(x)$ is the length of a particular geodesic from $h_i(x)$ to $u_i(x)$ and hence dominates the geodesic distance. Therefore we have

$$\int_B |h_i - u_i|^2 \leqslant C\ (E(u_i) - E_0).$$

Since $(E(u_i) - E_0) \to 0$, and $h_i$ converges to h while $u_i$ converges (weakly) to u it follows that u = h a.e. and hence u is smooth near x. A similar argument using the boundary a priori estimate, Corollary 2.8, shows that u is smooth in $\bar{M}$ and agrees with $\varphi$ on $\partial M$.

Finally we must show that u is homotopic to $\varphi$. One way to do this is to show that two smooth maps which have bounded energy and are $L^2$ close are actually homotopic (maps into a manifold of nonpositive curvature). A proof which is more in the present spirit is to modify $u_i$ by harmonic replacement on balls. Let $B_1, \ldots, B_p$ be a covering of $\bar{M}$ by balls. The argument given above shows that we can replace $u_i$ by a harmonic map $h_{i,1}$ in $B_1$, and the result, denoted $u_{i,1}$, becomes uniformly close to u in the interior of $B_1$. We then repeat the argument replacing $u_{i,1}$ on $B_2$ to get $u_{i,2}$ which by the above argument will be uniformly close to u in the interior of $B_1 \cup B_2$. Finally we get a map $u_{i,p}$ which is homotopic to $u_i$ and uniformly close to u. The maps $u_{i,p}$ and u are then homotopic and it follows that $u \sim \varphi$. This concludes the proof of Theorem 2.12.

## 3. Existence and Regularity of Weak Solutions

In this section we discuss existence questions for minima, and regularity results. We first prove a "higher regularity" result which says that Hölder continuous weak solutions are regular. This result is known to many experts, but only a special case [BG] appears explicitly in the literature. Note that a weak solution which has small oscillation has been shown to be Hölder continuous in [HKW], while a Lipschitz weak solution can easily be shown regular by standard linear elliptic theory. In our first lemma we bridge the gap between $C^\alpha$ and Lipschitz.

**Lemma 3.1.** *Let* $u \in C^\alpha(M,N) \cap L_1^2(M,N)$ *for some* $\alpha \in (0,1)$ *be a weakly harmonic map. Then* u *is a smooth map.*

**Proof.** We first show that u is locally Lipschitz. Let $x_0 \in M$, and $r_0 > 0$ such that the image of $B_{r_0}(x_0)$ under u lies in a uniformly Euclidean normal coordinate chart. Throughout this argument we work in this chart, so that u is thought of as a vector-valued function. We also choose normal coordinates $y^\beta$ in $B_{r_0}(x_0) \subset M$. Thus, from the Hölder continuity of u we have

$$g_{\beta\gamma}(y) = \delta_{\beta\gamma} + 0(\sigma^2)$$

(3.1)

$$h_{ij}(u(y)) = \delta_{ij} + 0(\sigma^{2\alpha})$$

for $|y| \leqslant \sigma$. The fact that u is weakly harmonic implies for any $\sigma \in (0,r_0)$

$$(3.2) \quad \left| \int_{B_\sigma(x_0)} \langle \nabla\eta, \nabla u \rangle \, dv_M \right| \leqslant C \sup_{B_\sigma(x_0)} |\eta| \int_{B_\sigma(x_0)} e(u) \, dv_M$$

for any $\eta \in C^\circ(\bar{B}_\sigma, \mathbb{R}^k) \cap L_1^2(B_\sigma, \mathbb{R}^k)$ with $\eta \equiv 0$ on $\partial B_\sigma$. Let v be the solution of the linear Laplace equation

$$\sum_\gamma \partial_\gamma \partial_\gamma v = 0 \text{ in } B_\sigma$$

$$v = u \text{ on } \partial B_\sigma$$

where $\partial_\gamma$ denotes the partial derivative with respect to $y^\gamma$. Since $|\partial v|^2 = \sum_{i,\gamma} \left(\frac{\partial v^i}{\partial y^\gamma}\right)^2$ is a (Euclidean) subharmonic function, we have the mean value inequality

$$(3.3) \quad \frac{d}{dr} \left[ ⨍_{B_r} |\partial v|^2 \, dy \right] \geqslant 0 \text{ for } 0 \leqslant r \leqslant \sigma$$

where $⨍_{B_r} f$ denotes the Euclidean average value of f. From (3.1) we have

$$\int_{B_{\frac{1}{2}\sigma}} |\partial u|^2 dy \leqslant (1 + C\sigma^\alpha) \int_{B_{\frac{1}{2}\sigma}} \langle \nabla u, \nabla u \rangle dv_M$$

$$= (1 + C\sigma^{\alpha})\left[\int_{B_{\frac{1}{2}\sigma}} \langle\nabla v, \nabla u\rangle dv_M + \int_{B_{\frac{1}{2}\sigma}} \langle\nabla(u-v), \nabla u\rangle dv_M\right]$$

From the Schwarz and arithmetic-geometric mean inequalities together with (3.1)

$$\left|\int_{B_{\frac{1}{2}\sigma}} \langle\nabla v, \nabla u\rangle \, dv_M\right| \leq (1 + C\sigma^{\alpha}) \left[\frac{1}{2} \int_{B_{\frac{1}{2}\sigma}} |\partial v|^2 \, dy\right.$$

$$\left. + \frac{1}{2} \int_{B_{\frac{1}{2}\sigma}} |\partial u|^2 \, dy\right]$$

Combining these inequalities with (3.2) (with $\eta = u - v$) we get

$$\fint_{B_{\frac{1}{2}\sigma}} |\partial u|^2 \, dy \leq (1 + C\sigma^{\alpha}) \fint_{B_{\frac{1}{2}\sigma}} |\partial v|^2 \, dy$$

$$+ C \sup_{B_\sigma} |u-v| \fint_{B_\sigma} e(u) \, dv_M$$

Applying (3.1), (3.3), and the Dirichlet minimizing property of v we get

$$\fint_{B_{\frac{1}{2}\sigma}} |\partial u|^2 dy \leq (1 + C\sigma^{\alpha}) \fint_{B_\sigma} |\partial v|^2 \, dy + C\sigma^{\alpha} \fint_{B_\sigma} |\partial u|^2 dy$$

$$\leq (1 + C\sigma^{\alpha}) \fint_{B_\sigma} |\partial u|^2 dy.$$

Setting $\sigma_i = r_0 2^{-i}$ we get

$$\int_{B_{\sigma_{i+1}}} |\partial u|^2\, dy \leqslant (1 + C\sigma_i^{\alpha}) \int_{B_{\sigma_i}} |\partial u|^2\, dy.$$

Upon iteration we get

$$\int_{B_{2^{-i}r_0}} |\partial u|^2\, dy \leqslant \bar{C} \int_{B_{r_0}} |\partial u|^2 \leqslant \bar{\bar{C}}$$

where $\bar{C} = \prod_{j=0}^{\infty} (1 + Cr_0^{\alpha}2^{-i\alpha}) < \infty$. Thus we see that for any $\sigma\epsilon(0,r_0)$ we have

$$\int_{B_{\sigma}(x_0)} |\partial u|^2\, dy \leqslant C.$$

Since $x_0$ was an arbitrary point we conclude that $|\partial u|^2$ is an $L^{\infty}$ function. Therefore it follows that u is locally Lipschitz. The equation for u then shows that $\Delta u$ is $L^{\infty}$ locally on M. Hence by linear elliptic regularity theory (see [M3,6.2.5]) we see that u is locally $L_2^p$ for $p < \infty$. Therefore we get $\Delta u \epsilon L_{1,loc}^p$ and hence $u \epsilon L_{3,loc}^p$. Repeating the argument shows that u is smooth. This proves Lemma 3.1.

<u>**Remark**</u> <u>**3.2**</u>. The smoothness assumptions necessary for the previous argument are that $g_{\beta\gamma}$ be Hölder continuous (or Dini continuous) and that $h_{ij}$ be $C^{1,\alpha}$. One can then show that u is $C^{1,\alpha}$.

<u>**Remark**</u> <u>**3.3**</u>. The proof of Lemma 3.1 can be modified in a straightforward way to prove boundary regularity assuming that u satisfies an appropriate Dirichlet boundary condition.

There are various regularity questions which arise in the harmonic map theory. For example, one can ask about regularity of weakly harmonic maps, stationary points, or energy minimizing maps. At present, there are no known regularity results for general weakly harmonic maps. The regularity of harmonic maps from a surface having finite energy and isolated singularities was shown in [SaU].

Our next theorem proves regularity of stationary maps defined on a surface. The special case when the map is weakly conformal was treated in [St].

__Theorem 3.2.__ *Suppose* $n = \dim M = 2$, *and* $u \in L_1^2(M,N)$ *is a stationary point for energy. Then* u *is a smooth harmonic map on the interior of* M.

__Proof.__ It suffices to prove that a stationary map defined on a unit disk D is smooth. From Lemma 1.1 we know that the function

$$\Phi(z) = \left(\left\| u_*\left(\frac{\partial}{\partial x}\right)\right\|^2 - \left\| u_*\left(\frac{\partial}{\partial y}\right)\right\|^2\right) - 2i \left\langle u_*\left(\frac{\partial}{\partial x}\right), u_*\left(\frac{\partial}{\partial y}\right)\right\rangle$$

is holomorphic in D. If $\Phi$ is identically zero, then u is weakly conformal and our theorem follows from [St]. Otherwise $\Phi$ has at most a finite number of zeroes $z_1, \ldots, z_s$ in $D_{1/2} = \{|z| < 1/2\}$. If we can prove that u is smooth on $D_{1/2} - \{z_1, \ldots, z_s\}$, then our theorem follows from [SaU]. In this way we are reduced to the case that $\Phi$ is nowhere zero in D. We deal with this case as follows. Let f(z) be an analytic function in D satisfying $f^2 = -\Phi$ in D. Let v(z) be the real harmonic function given by

$$v(z) = \frac{1}{2} \operatorname{Re} \int_0^z f(\zeta)\, d\zeta .$$

Thus v satisfies $\frac{\partial v}{\partial z} = \frac{1}{2} f$, $v(0) = 0$. Define a harmonic map $\tilde{u}: D \to N \times \mathbb{R}$ by $\tilde{u}(z) = (u(z), v(z))$, and observe that the Hopf differential of $\tilde{u}$ is given by

$$\tilde{\Phi}(z) = \Phi(z) + 4\left(\frac{\partial v}{\partial z}\right)^2(z) \equiv 0.$$

Thus $\tilde{u}$ is a weakly conformal harmonic map, and hence by the theorem of [St] is smooth. Therefore u is smooth and we have proved Theorem 3.2.

__Remark 3.3.__ No regularity theorem is known for stationary maps in dimension larger than two. Notice that Corollary 2.3 does assert partial regularity for maps which are locally (weakly) approximable by smooth harmonic maps. It is not at all clear that an arbitrary stationary map can be so approximated.

We now consider the case of minimizing maps. We first observe that the Dirichlet problem can be solved by the direct method to produce minimizing maps.

**Proposition 3.3.** *Let M be a compact manifold with nonempty boundary, and let $\varphi \in L_1^2(M,N)$ be a given map. There exists a map $u \in L_1^2(M,N)$ which has least energy over all maps $v \in L_1^2(M,N)$ which agree (in the $L_1^2$ sense) with $\varphi$ on $\partial M$. The regularity properties of u are described below.*

**Proof.** Let $\overset{\circ}{L}{}_1^2(M,\mathbb{R}^k)$ be the closure in $\overset{\circ}{L}{}_1^2$ norm of $C_0^\infty(M,\mathbb{R}^k)$, the smooth maps with compact support in the interior of M. We say that $v = \varphi$ on $\partial M$ if $v - \varphi \in \overset{\circ}{L}{}_1^2(M,\mathbb{R}^k)$. Let $\mathcal{F}_\varphi = \{v \in L_1^2(M,N): v = \varphi \text{ on } \partial M\}$, and let $E_\varphi$ be the infimum of the energy for maps in $\mathcal{F}_\varphi$. Since $\varphi \in \mathcal{F}_\varphi$ we have $E_\varphi \leq E(\varphi) < \infty$. Let $\{u_i\}$ be a sequence of maps in $\mathcal{F}_\varphi$ with

$$\lim_{i \to \infty} E(u_i) = E_\varphi.$$

By the weak compactness of bounded sets in $L_1^2(M,N)$, we can extract a subsequence of $\{u_i\}$ also denoted $\{u_i\}$ such that

$$u_i \rightharpoonup u \text{ (weak convergence)}$$

with $u \in L_1^2(M,N)$. Since $\overset{\circ}{L}{}_1^2(M,\mathbb{R}^k)$ is a closed linear subspace of $L_1^2(M,\mathbb{R}^k)$, Hilbert space theory tells us that $\overset{\circ}{L}{}_1^2(M,\mathbb{R}^k)$ is also weakly closed. Therefore, since $u_i - \varphi \in \overset{\circ}{L}{}_1^2(M,\mathbb{R}^k)$, we have that $u - \varphi \in \overset{\circ}{L}{}_1^2(M,\mathbb{R}^k)$ and hence $u \in \mathcal{F}_\varphi$. Also we have

$$E(u) = \int_M \|\nabla u\|^2 \, dv_M = \lim_{i \to \infty} \int_M \langle \nabla u_i, \nabla u \rangle \, dv_M$$

$$\leq \left[\varliminf_{i\to\infty} E(u_i)\right]^{1/2} (E(u))^{1/2}$$

by the Schwarz inequality and weak convergence of $u_i$ to u. Therefore we have $E(u) \leq E_\varphi$, and the map u minimizes energy over all maps in $\mathcal{F}_\varphi$. This proves Proposition 3.3.

If in addition to specifying the boundary values of a minimizing map, we attempt to specify its homotopy class, we encounter the difficulty that the homotopy class is not generally a weakly closed condition in $L_1^2$. In fact, we mention the following recent result of Brian White [W]. Suppose for simplicity that $\partial M = \emptyset$, and let $\varphi \in C^\infty(M,N)$ be a given map. Let $E_\varphi$ denote the infimum of energy taken over all smooth maps v which are homotopic to $\varphi$. An old observation of Morrey is that if $M = N = S^n$, $n > 2$, and $\varphi$ is the identity map, then $E_\varphi = 0$. The following is a result of [W].

__Proposition 3.4.__ $E_\varphi = 0$ *if and only if the homomorphism induced by* $\varphi$ *on both* $\pi_1$ *and* $\pi_2$ *is zero.*

Several authors ([L], [SaU], [ScY2]) have proven existence theorems for minimizing maps of surfaces with specified behavior on $\pi_1$. Precisely, we consider the following definition.

__Definition 3.5.__ Two continuous maps $v, \varphi$ are __$\pi_1$-equivalent__ if the homomorphisms $v_*:\pi_1(M,x_0) \to \pi_1(N,v(x_0))$ and $\varphi_*:\pi_1(M,x_0) \to \pi_1(N,\varphi(x_0))$ are conjugate along some path $\sigma$ from $v(x_0)$ to $\varphi(x_0)$; i.e. $v_* = \sigma^{-1}\varphi_*\sigma$.

It is easy to see that continuous maps which are $L_1^2$ close (even weakly) are $\pi_1$-equivalent. Thus we see that if the continuous maps were dense in $L_1^2(M,N)$ we could associate with any map $u \in L_1^2(M,N)$ a $\pi_1$-equivalence class of continuous maps (maps which approximate u sufficiently well). We could then hope to preserve this $\pi_1$-equivalence class under weak convergence, and hence construct a map u of least energy among all maps which are $\pi_1$-equivalent to a given map $\varphi$. Simple examples show (see [SU2]) that the continuous maps are not generally dense in $L_1^2(M,N)$; however, if dim M = 2 it is shown in [SU2] that $C^\infty(M,N)$ is dense in $L_1^2(M,N)$. This can be used to simplify the proof [ScY2] of the following result.

**Theorem 3.6.** *Suppose* $\dim M = 2$, *and* $\varphi \in C^\infty(M,N)$. *There exists a harmonic map* $u \in C^\infty(M,N)$ *which is* $\pi_1$*-equivalent to* $\varphi$, *and which has least energy over all* $L_1^2$ *maps* v *which are* $\pi_1$*-equivalent to* $\varphi$.

**Proof.** Let $\mathcal{F}_\varphi = \{v \in L_1^2(M,N):$ v is $\pi_1$-equivalent to $\varphi\}$ where the meaning of the $\pi_1$-equivalence class of $v \in L_1^2(M,N)$ is the equivalence class of any smooth approximating map. It is then easy to show that bounded subsets of $\mathcal{F}_\varphi$ are weakly closed (hence weakly compact). Thus by the direct method we can construct $u \in \mathcal{F}_\varphi$ of least energy. The smoothness of u then follows by adapting a theorem of Morrey [M2]. This completes a sketch of the proof of Theorem 3.6.

The generalization of the above result to higher dimensions is more complicated. It has recently been carried out by B. White [W]. The idea is to prove that a map $u \in L_1^2(M,N)$ can be approximated by a map $\hat{u}$ which is continuous on a suitably chosen 2-skeleton $Z^2$ of M. Since $\pi_1(Z^2)$ is isomorphic under the inclusion map to $\pi_1(M)$, the map $\hat{u}$ defines a $\pi_1$-equivalence class. Two such approximations can be shown to be $\pi_1$-equivalent. It is then shown that the energy can be minimized in a $\pi_1$-equivalence class.

**Theorem 3.7.** [W] *Let* $\varphi \in L_1^2(M,N)$, *and denote by* $\mathcal{F}_\varphi$ *the set of* $L_1^2(M,N)$ *maps* v *which are* $\pi_1$*-equivalent to* $\varphi$ *(as described above). There exists* $u \in \mathcal{F}_\varphi$ *of least energy over all maps in* $\mathcal{F}_\varphi$. *The regularity properties of* u *are described below.*

We now describe the regularity properties of energy minimizing maps, in particular those which are constructed in Proposition 3.3 and Theorem 3.7. These results are proved in [SU1], [SU2]. First note that any smooth harmonic map $u: S^n \to N$ for $n \geq 2$ gives rise to a weakly harmonic (in fact, stationary) map $\bar{u}: \mathbb{R}^{n+1} \to N$ with an isolated singularity at 0 by setting

$$\bar{u}(x) = u\left(\frac{x}{|x|}\right) \text{ for } x \neq 0.$$

Such a map $\bar{u}$ is called a tangent map. If $\bar{u}$ minimizes energy on compact subsets of $\mathbb{R}^{n+1}$, then $\bar{u}$ is a minimizing tangent map. Thus a minimizing tangent map has the property that for any bounded domain $\Omega \subset \mathbb{R}^{n+1}$ and any map $v \in L_1^2(\Omega,N)$ with $v = \bar{u}$ on $\partial\Omega$, we have $E_\Omega(\bar{u}) \leq E_\Omega(v)$. (Actually, it suffices to take $\Omega$ to be the unit ball in $\mathbb{R}^{n+1}$ by homogeneity of $\bar{u}$.) Certain manifolds admit no nontrivial tangent maps; e.g. any manifold N whose universal cover carries a strictly convex function. On the other hand, a theorem of [SaU] shows that any compact manifold N whose universal cover is not contractible contains a harmonic $S^2$, i.e. a nonconstant smooth harmonic $u:S^2 \to N$. Thus such manifolds always contain tangent maps. The condition that a tangent map $\bar{u}$ be minimizing is clearly much stronger, and one expects such $\bar{u}$ to be more rare (see [SU3] for some evidence). The tangent maps indicate that harmonic maps will in general have singularities, so in order to develop a regularity theory one must allow for this possibility. We say that a point x in the interior of M is a regular point for a map u if u is $C^1$ in a neighborhood of x. The set $\mathcal{R}$ of regular points is then a (possibly empty) open subset of Int(M). The complement of $\mathcal{R}$ is denoted $\mathcal{S}$, and is the singular set of u. That $\mathcal{S}$ may be nonempty for a harmonic map is shown by the tangent maps $\bar{u}$. The Hausdorff dimension, dim $\mathcal{S}$, of a set $\mathcal{S}$ is the greatest lower bound of real numbers $t \geq 0$ such that $\mathcal{S}$ can be covered by balls $\{B_{r_i}(x_i)\}$ with

$\sum_i r_i^t < \varepsilon$ for any preassigned positive number $\varepsilon$. It is clear that

for a set $\mathcal{S} \subset M^n$ we have $0 \leq \dim \mathcal{S} \leq n$. The interior regularity theorem of [SU1] is as follows.

**Theorem 3.8.** *Suppose* $u \in L_1^2(M,N)$ *is a minimizing map. Then* $\dim \mathcal{S} \leq n-3$. *In particular* $\mathcal{S} = \emptyset$ *if* $n = 2$. *If* $n = 3$ *then* $\mathcal{S}$ *is a discrete set of points. Moreover, if every minimizing tangent map from* $\mathbb{R}^{p+1}$ *into* N *is constant for* $2 \leq p \leq q$, *then* $\dim \mathcal{S} \leq n-q-2$ ($\mathcal{S} = \emptyset$ *if* $n < q + 2$). *If* $n = q + 2$, *then* $\mathcal{S}$ *is discrete.*

**Remark 3.9.** The case n = 2 of Theorem 3.8 was proved by

Morrey [M2]. A partial regularity result for systems of functions minimizing variational integrals was recently obtained by Giaquinta-Guisti [GG1], [GG2]. The case n = 2 of their theorem had been done in an earlier paper of Morrey [M1]. See the next section for a more detailed discussion of the relations of the harmonic map problem with elliptic systems.

**Remark 3.10.** In [SU2], the question of boundary regularity has been considered. It is shown that minimizing maps are regular in a neighborhood of $\partial M$ provided a Dirichlet boundary condition is imposed on the map. The ideas of [GG1] have been extended to the boundary by Jost-Meier [JM] to assert boundary regularity for elliptic systems minimizing energy-type integrals.

## 4. Relationship to Nonlinear Elliptic Systems

If one considers a map u whose image is assumed to lie in a uniformly Euclidean coordinate chart, then the energy functional becomes

$$E(u) = \int_M \sum_{\alpha,\beta,i,j} g^{\alpha\beta}(x)\, h_{ij}(u(x)) \frac{\partial u^i}{\partial x^\alpha} \frac{\partial u^j}{\partial x^\beta} (g)^{1/2}\, dx.$$

This is a special case of a functional of the form

$$(4.1)\qquad \underline{\underline{F}}(u) = \int_\Omega F(x,u(x),\nabla u(x))\, dx$$

for a vector valued function $u:\Omega \to \mathbb{R}^k$ where $\Omega$ is a domain in $\mathbb{R}^n$. Here we have the integrand F(x,u,P) is a real valued function

$$F:\Omega \times \mathbb{R}^k \times \mathbb{R}^{nk} \to \mathbb{R}$$

where $P = (P^i_\alpha) \in \mathbb{R}^{nk}$. It is natural to consider vector functions which are critical points of such functionals $\underline{\underline{F}}$. For a recent account of results in this direction see the book of M. Giaquinta [G]. From the point of view of differential geometry, the restriction that the image of a map lie in a coordinate ball is an unpleasant one. In particular, many of the PDE results rely on the linear structure of the target, and hence do not capture the more interesting global aspects of the harmonic map problem. (Note that the local estimates and regularity results are tied to the global geometry of the target because the oscillation of the map can be arbitrarily large locally.) An interesting generalization of the harmonic map problem would be to study critical points of the functional of (4.1) (with suitable ellipticity and bounds assumed) in the space $L_1^2(\Omega,N)$, i.e. with a pointwise constraint imposed on the maps. The regularity results of [SU1] require the monotonicity inequality (1.4) which probably isn't true for a large class of functionals $\underline{E}$. It is not known to the author whether minima of more general elliptic functionals in $L_1^2(\Omega,N)$ have any partial regularity properties.

Next we record an interesting example of Frehse [F] which gives a smooth elliptic system in two variables (with a structure reminiscent of the harmonic map system) having a singular weak solution. The system is given by

$$(4.2)\qquad \begin{aligned} &\Delta u^1 + \frac{2(u^1+u^2)}{1+|u|^2}|\nabla u|^2 = 0 \\ &\Delta u^2 + \frac{2(u^2-u^1)}{1+|u|^2}|\nabla u|^2 = 0. \end{aligned}$$

The functions $u^1(x) = \sin\log\log\frac{1}{|x|}$, $u^2(x) = \cos\log\log\frac{1}{|x|}$ provide a bounded $L_1^2$ weak solution with an isolated singularity at $x = 0$. Whether a singular weak solution of a harmonic map equation exists in two dimensions is unknown. The result of [SaU] and Theorem 3.2 provide evidence against the existence of such a weak solution. We also show here that a priori estimates for smooth solutions (generalizing Theorem 2.2) do hold for general equations (including (4.2)). This shows that the singular weak solution of (4.2) cannot be an $L_1^2$ limit of smooth solutions. Let $\mathscr{L}$ be an elliptic operator of the form $\mathscr{L} = (\mathscr{L}_1, \ldots, \mathscr{L}_k)^t$

$$\mathscr{L}_j U = \sum_{i=1}^{k}\sum_{\alpha,\beta=1}^{2}\frac{\partial}{\partial x^\alpha}\left(a_{ij}^{\alpha\beta}\frac{\partial u^i}{\partial x^\beta}\right) + \sum_{i=1}^{k}\sum_{\alpha=1}^{2} b_{ij}^{\alpha}\frac{\partial u^i}{\partial x^\alpha} + \sum_{i=1}^{k} c_{ij}u^i$$

where $U = (u^1, \ldots, u^k)^t$. Assume a, b, c are Hölder continuous and $\mathcal{L}$ is elliptic in the sense that

$$\det(\alpha_{ij}(x,\lambda)) \neq 0, \qquad \alpha_{ij} = \sum_{\alpha,\beta=1}^{2} a_{i\,j}^{\alpha\beta}(x)\,\lambda_\alpha\lambda_\beta$$

for any $\lambda \in \mathbb{R}^2$, $\lambda \neq 0$. The following elliptic estimate then holds (see [M3,6.2.6])

$$(4.3) \qquad \|U\|_{C^{1,\alpha}(D_{1/2})} \leqslant C(\|\mathcal{L}U\|_{L^\infty(D_1)} + \|U\|_{L^\infty(D_1)})$$

where $U \in C^2(D_1)$, $D_r = \{x: |x|<r\}$, $\mathcal{L}U \in L^\infty(D_1)$. The following a priori estimate is based on (4.3) together with a scaling argument. Consider a nonlinear system of the form

$$\mathcal{L}U = A(x,U,\nabla U)$$

where A is a locally $L^\infty$ function on $\mathbb{R}^2 \times \mathbb{R}^k \times \mathbb{R}^{2k}$ which satisfies

$$(4.4) \qquad |A(x,U,P)| \leqslant C_1|P|^2 + C_2$$

for $|x| + |U| \leqslant T$ with $C_1, C_2$ depending on T.

**Theorem 4.1.** *Suppose* U *is a* $C^2$ *solution of* $\mathcal{L}U = A(x,U,\nabla U)$ *on* $D_1$. *There exists* $\varepsilon_0 > 0$ *depending only on* $\mathcal{L}$, A, *and* $\sup_{D_1}|U|$ *such that if*

$$\int_{D_1} |\nabla U|^2\,dx \leqslant \varepsilon_0,$$

*then we have the estimate*

$$\sup_{D_{1/2}} |\nabla U|^2 \leqslant C \int_{D_1} |\nabla U|^2\,dx.$$

*where* C *depends on* $\mathscr{L}$, A, *and* $\sup_{D_1} |U|$.

__Proof.__ We recall the proof of Theorem 2.2. By the same rescaling argument we get a solution V on $D_{r_0}$ of an equation

(4.5) $\mathscr{L}'V = A'(x,V,\nabla V)$

where $|\nabla V|^2 \leqslant 4$ on $D_{r_0}$, and $|\nabla V|^2(0) = 1$. Recall that $r_0 = (e_0)^{1/2}\,\rho_0$ where

$$e_0 = \sup_{D_{\sigma_0}} |\nabla U|^2, \qquad \rho_0 = \frac{1}{2}(1 - \sigma_0)$$

and $\sigma_0 \in (0,1)$ satisfies

$$(1 - \sigma_0)^2 \sup_{D_{\sigma_0}} |\nabla U|^2 = \max_{0<\sigma<1} (1 - \sigma)^2 \sup_{D_\sigma} |\nabla U|^2.$$

Since $\mathscr{L}'$, A' are simply rescaled versions of $\mathscr{L}$, A we see from (4.4) that A'(x,U,P) is still of quadratic growth in P, and hence we have

$$\sup_{D_{r_0}} |A'(x,V,\nabla V)| \leqslant C.$$

If $r_0 \geqslant 1$, then we can apply the estimate (4.3) to $\mathscr{L}'$ (note that the Hölder norms of a, b, c decrease upon rescaling) to conclude

(4.6) $\|V\|_{C^{1,\alpha}(D_{1/2})} \leqslant C$

where C depends only on $\mathscr{L}$, A, and $\sup_{D_1} |U|$. On the other hand we know from the scale invariance of the Dirichlet integral that

$$(4.7) \quad \int_{D_1} |\nabla V|^2 \leqslant \int_{D_1} |\nabla U|^2 < \varepsilon_0.$$

Since $|\nabla V|(0) = 1$, we see that (4.6) and (4.7) are contradictory if $\varepsilon_0$ is small. Therefore we get $r_0 \leqslant 1$, and hence we have

$$\max_{0<\sigma<1} (1 - \sigma)^2 \, |\nabla U|^2 \leqslant 4.$$

In particular, $|\nabla U|$ is bounded on $D_{3/4}$, and the conclusion of the theorem now follows from (4.3) and (4.4).

## References


[BG] H. J. Borchers and W. J. Garber, Analyticity of solutions of the O(N) non-linear $\sigma$-model, Comm. Math. Phys. 71(1980), 299-309.

[CG] S. S. Chern and S. Goldberg, On the volume-decreasing property of a class of real harmonic mappings, Amer. J. Math. 97(1975), 133-147.

[ES] J. Eells and J. Sampson, Harmonic mappings of Riemannian manifolds, Amer. J. Math. 86(1964), 109-160.

[EL] J. Eells and L. Lemaire, A report on harmonic maps, Bull. London Math. Soc. 10(1978), 1-68.

[F] J. Frehse, A discontinuous solution to a mildly nonlinear elliptic system, Math. Z. 134(1973), 229-230.

[G] M. Giaquinta, Multiple integrals in the calculus of variations and nonlinear elliptic systems, Annals of Math. Studies 105, 1983.

[GG1] M. Giaquinta and E. Giusti, On the regularity of the minima of variational integrals, Acta. Math. 148(1982), 31-46.

[GG2] M. Giaquinta and E. Giusti, The singular set of the minima of certain quadratic functionals, to appear in Ann. Sc. Norm. Sup. Pisa.

[GH] M. Giaquinta and S. Hildebrandt, A priori estimates for harmonic mappings, J. Reine Angew. Math. 336(1982), 124-164.

[Hm] R. Hamilton, Harmonic maps of manifolds with boundary, Lecture notes 471, Springer 1975.

[Hr] P. Hartman, On homotopic harmonic maps, Can. J. Math. 19(1967), 673-687.

[HKW] S. Hildebrandt, H. Kaul, and K. O. Widman, An existence theorem for harmonic mappings of Riemannian manifolds, Acta Math. 138(1977), 1-16.

[HW] S. Hildebrandt and K. O. Widman, On the Hölder continuity of weak solutions of quasilinear elliptic systems of second order, Ann. Sc. Norm. Sup. Pisa IV(1977), 145-178.

[JM] J. Jost and M. Meier, Boundary regularity for minima of certain quadratic functionals, Math. Ann. 262(1983), 549-561.

[L] L. Lemaire, Applications harmoniques de surfaces Riemanniennes, J. Diff. Geom. 13(1978), 51-78.

[M1] C. B. Morrey, On the solutions of quasilinear elliptic partial differential equations, Trans. A.M.S. 43(1938), 126-166.

[M2] C. B. Morrey, The problem of Plateau on a Riemannian manifold, Ann. of Math. 49(1948), 807-851.

[M3] C. B. Morrey, Multiple integrals in the calculus of variations, Springer-Verlag, New York, 1966.

[P] P. Price, A monotonicity formula for Yang-Mills fields, Manuscripta Math. 43(1983), 131-166.

[ScY1] R. Schoen and S. T. Yau, Compact group actions and the topology of manifolds with non-positive curvature, Topology 18(1979), 361-380.

[ScY2] R. Schoen and S. T. Yau, Existence of incompressible minimal surfaces and the topology of three dimensional manifolds with non-negative scalar curvature, Ann. of Math. 110(1979), 127-142.

[S] Y. T. Siu, The complex analyticity of harmonic maps and the strong rigidity of compact Kähler manifolds, Ann. of Math. 112(1980), 73-111.

[SiY] Y. T. Siu and S. T. Yau, Compact Kähler manifolds of positive bisectional curvature, Invent. Math. 59(1980), 189-204.

[SaU] J. Sacks and K. Uhlenbeck, The existence of minimal immersions of 2-spheres, Ann. of Math. 113(1981), 1-24.

[SU1] R. Schoen and K. Uhlenbeck, A regularity theory for harmonic maps, J. Diff. Geom. 17(1982), 307-335.

[SU2] R. Schoen and K. Uhlenbeck, Boundary regularity and the Dirichlet problem for harmonic maps, J. Diff. Geom. 18(1983), 253-268.

[SU3] R. Schoen and K. Unlenbeck, Regularity of minimizing harmonic maps into the sphere, to appear in Invent. Math.

[St] M. Struwe, preprint.

[W] B. White, Homotopy classes in Sobolev spaces of mappings, preprint.

# EQUATIONS OF PLASMA PHYSICS

**Alan Weinstein**

**Notes by Stephen Omohundro**

## (a) Plasmas

A plasma is a gas of charged particles under conditions where collective electromagnetic interactions dominate over interactions between individual particles. Plasmas have been called the fourth state of matter [1]. As one adds heat to a solid, it undergoes a phase transition (melting) to become a liquid. More heat causes the liquid to boil into a gas. Adding still more energy causes the gas to ionize (i.e. some of the negative electrons become dissociated from their gas atoms, leaving positively charged ions). Above 100,000°K, most matter ionizes into a plasma. While the earth is a relatively plasma-free bubble (aside from fluorescent lights, lightning discharges, and magnetic fusion energy experiments) 99.9% of the universe is in the plasma state (e.g. stars and most of interstellar space).

Though a plasma is a gas, the dominance of collective interactions makes the behavior quite different from that described by gas dynamics. The so-called plasma parameter $g = \left[\frac{8\pi e^2}{kT}\right]^{3/2}\sqrt{n}$ (e is the charge of an electron, n is the electron density and kT is the temperature in energy units) must be much less than unity for this "plasma approximation" to be valid. In this case, it is appropriate to replace an exact description of the position and velocity of every particle by a density $f_\alpha(x,v,t)\, dx\, dv$ in position-velocity phase space for every "species" of particle $\alpha$ (eg. electrons, different types of ions). This gives the number of particles of a given species per unit volume at a particular value of $(x,v) \in \mathbb{R}^6$. The exact description would make $f_\alpha$ a sum of $\delta$-functions, one for each particle, but the collective nature of the plasma suggests that it is a useful idealization to assume the $f_\alpha$'s to be smooth. Notice that in hydrodynamics one usually assumes the velocity to be a function of position, whereas here we may have particles of many different velocities at each point in space.

Particles move in phase space for two reasons: they change their spatial position x if they have a non-zero velocity, and they change their velocity v if they feel a non-zero force. Here the only force we will consider is the electromagnetic Lorentz force $e_\alpha(E+\frac{v}{c}\times B)$ due to the electric field E and magnetic field B. These are vector fields on spatial $\mathbb{R}^3$. $e_\alpha$ is the charge of a particle of species $\alpha$, and c is the speed of light (which appears because we are using Gaussian units) [2].

If the E and B fields were due to external agents, and the particles did not affect one another, then the evolution of the distribution $f_\alpha$ would be just transport along the flow in phase space generated by the Liouville equation:

$$\frac{\partial f_\alpha}{\partial t} + v\cdot\frac{\partial f_\alpha}{\partial x} + \frac{e_\alpha}{m_\alpha}\left[E + \frac{v}{c}\times B\right]\cdot\frac{\partial f_\alpha}{\partial v} = 0,$$

where $m_\alpha$ is the mass of a particle of species $\alpha$, $\frac{\partial f_\alpha}{\partial x}$ is the gradient of $f_\alpha$ in the x directions, and $\frac{\partial f_\alpha}{\partial v}$ is the v-gradient. Here, E and B are prescribed vector-valued functions of space and time; given initial values for the $f_\alpha$, one seeks their evolution in time.

The complexity of real plasma behavior is due to the fact that the E and B fields are affected by the plasma particles themselves. One thinks of the plasma as having a charge density

$$\rho_f(x) = \Sigma_\alpha e_\alpha \int f_\alpha(x,v)dv$$

and a current density

$$j_f(x) = \Sigma_\alpha e_\alpha \int f_\alpha(x,v)v dv.$$

These act as sources for the electromagnetic field, which evolves according to Maxwell's equations. The entire coupled set is called the Maxwell-Vlasov equations [3]:

$$\left.\frac{\partial f_\alpha}{\partial t} + v \cdot \frac{\partial f_\alpha}{\partial x} + \frac{e_\alpha}{m_\alpha}\left[E + \frac{v}{c} \times B\right] \cdot \frac{\partial f_\alpha}{\partial v} = 0 \quad \right\} \text{Vlasov}$$

$$\frac{1}{c}\frac{\partial E}{\partial t} = \nabla \times B - \left[\Sigma_\alpha e_\alpha \int f_\alpha(x,v)\, v\, dv\right]\frac{4\pi}{c}$$

$$\frac{1}{c}\frac{\partial B}{\partial t} = -\nabla \times E$$

$$\text{Maxwell}$$

$$\nabla \cdot B = 0$$

$$\nabla \cdot E = 4\pi\, \Sigma_\alpha e_\alpha \int f_\alpha(x,v)\, dv$$

The electromagnetic field here is due to the smoothed particle distribution. If one wanted to include some particle discreteness effects, the zero in the Vlasov equation could be replaced by a term giving the change in $f_\alpha$ due to particle "collisions" (which are really just electromagnetic interactions that we ignored when we smoothed $f_\alpha$). In this case the equation is known as the Boltzmann equation and the typical effect is to cause distributions to relax to "Maxwellian" form $f(x,v) \propto e^{-mv^2/2kT}$ at each point x.

**(b) Simplifications**

We may treat the Maxwell-Vlasov equations as an initial value problem where we are given initial conditions $f_\alpha(x,v)$, $E(x)$, and $B(x)$ consistent with the constraints, and we wish to solve for the time evolution. As yet, only a little is known rigorously about the existence, uniqueness and qualitative behavior of the solutions to these equations. To help with analysis, one looks at a series of simplified sets of equations which are expected to realistically model the physical behavior in restricted parameter regimes. Understanding the relations between the animals in this zoo of simplifications is itself a challenging mathematical problem.

I. If we are interested in slow electrostatic phenomena we may take the limit $c \longrightarrow \infty$. This eliminates magnetic effects and makes the electro-magnetic field a slave to the particles, with no independent evolution of its own. The resulting system of equations is known as the Poisson-Vlasov equations:

$$\frac{\partial f_\alpha}{\partial t} + v \cdot \frac{\partial f_\alpha}{\partial x} + \frac{e_\alpha E}{m_\alpha} \cdot \frac{\partial f_\alpha}{\partial v} = 0$$

$$\text{div } E = \rho^f, \qquad \text{curl } E = 0$$

Poisson's name is used here because curl $E = 0$ implies that $E$ arises from a potential $V$ satisfying Poisson's equation $\Delta V \propto \rho^f$. This same set of equations describes the evolution of a "gas" of massive bodies, like stars, under the influence of Newtonian gravitational forces. In this context something about the existence and uniqueness of solutions is known. In 4 spatial dimensions there is collapse for Newtonian interactions. In 2 dimensions the solution is fine for all time if the initial conditions are compactly supported [4]. Not much is known about the physically relevant 3 dimensional case.

II. If we are dealing with a cold plasma, then there is not much dispersion in the velocity distribution function at each point. In this case it is reasonable to restrict attention to distributions that are δ-functions in the velocity direction:

$$f_\alpha(x,v,t) = \delta(v - v_\alpha(x,t))\, \rho_\alpha(x,t)$$

The evolution here is closely related to compressible fluid equations.

III. In both the stellar dynamics and the electrostatic plasma situations, people have studied so called "water-bag" models [5] where the distribution function is 1 in a region of x,v space and 0 outside it:

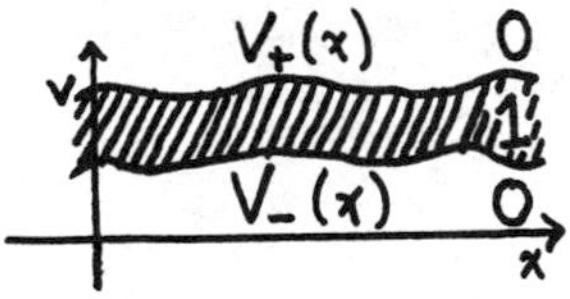


If we let this distribution evolve via the Liouville equations, it remains a water bag, but the bag might try to curl over itself:

representing the formation of a shock wave. Water-bag models are good for numerical studies. For example, initial conditions representing 2 streams with different velocities

typically develop "fingers"

which sprout "tendrils" and get tangled up:

It is a challenge to understand this behavior, perhaps utilizing strange attractors.

IV. If the magnetic field is large and primarily external, then the particles like to wind around the field lines in tight little helices (because of the v × B force): 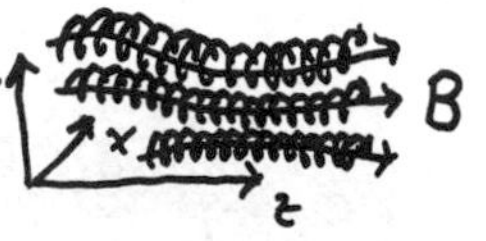

We can write the equations of motion for the so-called "guiding center" of these little loops. Considering a distribution function of guiding centers, we get a guiding center plasma. In the presence of external forces or gradients in the magnetic field, the guiding centers "drift" according to specifiable laws. For example, in the presence of an electric field, we get drifts in the direction of E × B:

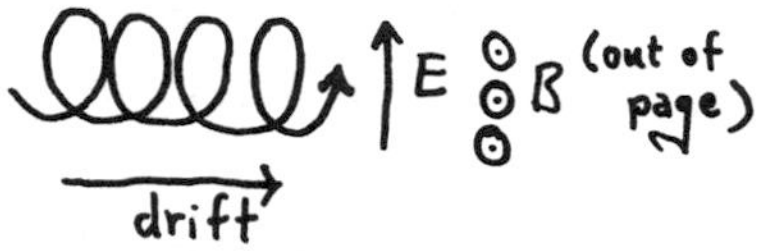

If the equipotentials enclose bounded regions, then the drifts can flow in loops: 

. Restricting to x,y and calling the guiding center density $\rho(x,y)$, the evolution in a constant magnetic field $B_z$ and an electric field due to the particles and a uniform neutralizing background is given by $\rho_t = \{\Delta^{-1}(\rho - 1), \rho\}$ where $\{a,b\} = a_x b_y - a_y b_x$. This equation is precisely the same as that which describes vorticity evolution in 2 dimensional fluids. Much of the analysis of fairly artificial vorticity examples becomes quite realistic in this context. For example a vortex blob (vorticity 1 in a region, 0 outside it) 

is very difficult to approximate experimentally, but a guiding center blob is just a beam of particles. One finds a variety of complicated behaviors in the evolution of these blobs [6], including stationary patterns, fingers, and oscillations, similar to those seen for water bags. As another example, point vortices 

correspond to a collection of discrete beams.

V. The previous four simplifications were subsystems of the Maxwell-Vlasov system. The last simplification is a quotient instead. Instead of keeping track of the whole distribution function, $f_\alpha(x,v,t)$, we focus attention on the low order velocity moments, which turn out to be physically meaningful quantities. The zeroth moment is proportional to the charge density $\rho_\alpha(x,t) = e_\alpha \int f_\alpha(x,v,t)\, dv$ or mass density. The local average velocity is related to the first order moment by $\bar{v}_\alpha(x,t) = \dfrac{\int f_\alpha(x,v,t)\, v\, dv}{\int f_\alpha(x,v,t)\, dv}$. The tensor density $p^\alpha_{ij}$ is related to the second order moments by:

$$p^\alpha_{ij}(x,t) = \int (v_i - \bar{v}_i)(v_j - \bar{v}_j)\, f_\alpha(x,v,t)\, dv.$$

We have a map $f \longmapsto (\rho,v)$, and one would hope that the flow in f-space "covers" the flow in $\rho$,v-space. Unfortunately the $(\rho,v)$ evolution equations contain $p_{ij}$, and the $p_{ij}$ evolution depends on higher moments in a never-ending heirarchy. In practice, one truncates this hierarchy and makes the assumption that $p_{ij}$ is a given function of $\rho$. Exactly this assumption is made in obtaining the equations for compressible fluids, and so the reduced equations are called the 2-fluid model for the plasma.

It is a general problem to understand the approximation processes here, perhaps in the way that the relationship between compressible and incompressible fluid descriptions was elucidated in Majda's talk in this seminar. An understanding of how certain properties are preserved has come from a geometric picture that has developed in the last 2-3 years, based on one which Arnold painted for inviscid fluids about 15 years ago [7]. We describe this development in the following section.

## (c) Hamiltonian Structure

Poisson manifolds are a generalization of symplectic manifolds, which have been used in geometric mechanics for about 30 years. A Poisson manifold is a manifold M with a Poisson bracket $\{,\}$ defined so as to make $C^{\infty}(M)$ a Lie algebra and additionally to satisfy a Leibniz rule: $\{fg,h\} = f\{g,h\} + g\{f,h\}$. More geometrically, this gives rise to a bilinear form B on the cotangent bundle, defined by $B(df,dg) = \{f,g\}$. (Many aspects of this structure were discovered by Lie [8] and were recently rediscovered by various people [9] with no reference to Lie.) B gives rise to a map $\tilde{B}: T^*M \longrightarrow TM$. If B is non-degenerate, we may invert this map to get a symplectic 2-form. Poisson structures are more interesting because they <u>can</u> be degenerate.

The range of $\tilde{B}$ is a linear subspace of each fibre in TM. Thus it is like a distribution, but the dimension jumps as we move around in M. The Jacobi identity for $\{,\}$ implies that the Frobenius integrability condition is satisfied, thus, in so-called <u>regular</u> regions where the range of $\tilde{B}$ is of constant dimension, M is foliated by smooth leaves. Since $\{,\}$ restricted to these leaves is non-degenerate, they have a natural symplectic structure. In fact, the symplectic leaves exist even at non-regular points, so Poisson manifolds are a promising model for the Hamiltonian formulation of physical situations with parameters, in which the number of variables changes at certain limiting values (e.g. $c \longrightarrow \infty$).

Poisson manifolds are the home of Hamiltonian systems. If $H: M \longrightarrow \mathbb{R}$ is a Hamiltonian function, then we call $\widetilde{B}(dH) \equiv \xi_H$ the Hamiltonian vector field associated to it. For an arbitrary function F we have $\xi_H \cdot F = \{F,H\}$ so $\dot{F} = \{F,H\}$ defines the flow of $\xi_H$. In coordinates $x_i$ this says

$$\dot{x}_i = \{x_i,H\} = \sum_{j=1}^{\dim M} \{x_i,x_j\} \frac{\partial H}{\partial x_j}.$$

This form of Hamilton's equations is due to Poisson. Note that all Hamiltonian vector fields are tangent to symplectic leaves.

In this framework, we may formulate the stability criterion of Lagrange and Arnold [7]. If $p \in M$ is an equilibrium point for $\xi_H$, then dH restricted to the symplectic leaf containing p must vanish at p. If $d^2H$ restricted to this leaf is definite, then p is a stable equilibrium point for the flow restricted to the leaf. If in addition p is in the regular region, then stability holds for the entire flow.

There is a natural Poisson bracket called the Lie-Poisson bracket defined on $g^*$, the dual of a Lie algebra g. If g is the Lie algebra of a group G, then we may identify left invariant functions on $T^*G$ with functions on $g^*$. $T^*G$ has the natural canonical Poisson bracket which is preserved by left translations, so the space of left invariant functions is closed under the operation of Poisson brackets. Identifying this subspace with $C^\infty(g^*)$, we obtain the Lie-Poisson bracket:

$$\{F,G\}(\mu) = \left\langle \mu, \left[\frac{\delta F}{\delta \mu}, \frac{\delta G}{\delta \mu}\right] \right\rangle$$

where $\mu \in g^*$, $F,G \in C^\infty(g^*)$, $\frac{\delta F}{\delta \mu}, \frac{\delta G}{\delta \mu} \in g$ and $\langle\ ,\ \rangle$ is the natural pairing between $g^*$ and g. In terms of the structure constants $c_{ijk}$ for g this is:

$$\{F,G\}(\mu) = \sum_{ijk} c_{ijk}\, \mu_i \frac{\partial F}{\partial \mu_j} \frac{\partial G}{\partial \mu_k}.$$

The symplectic leaves in $g^*$ of this Poisson bracket are exactly the orbits of the coadjoint representation of G in $g^*$. This fact was "discovered" by Kirillov and others ~ 1960, but Lie had the result in 1890. Lie also showed that any homogeneous symplectic manifold is one of these leaves, thus "prediscovering" further results of the 1960's.

Poisson bracket preserving maps $J: M \longrightarrow g^*$ from a Poisson manifold to the dual of a Lie algebra are called momentum mappings. By evaluation, such a map gives rise to a map $g \longrightarrow C^\infty(M)$. Viewing the image of an element of g as a Hamiltonian function, such maps are associated with Poisson bracket preserving actions of G on M. Usually one goes the other way. Start with a G action on M, get the associated momentum mapping, and then notice that it is G equivariant. If M is a homogeneous symplectic manifold, then the momentum mapping must be a covering onto a coadjoint orbit.

All of the plasma equations mentioned above have Poisson structures. A structure for the Maxwell-Vlasov equations was found by Morrison [10]. An error [11] was corrected by Marsden and Weinstein [12].

For the Poisson-Vlasov equations, we consider the infinite-dimensional Lie algebra $g = C^\infty(\mathbb{R}^6)$ of functions on $\mathbb{R}^6 = (x,v)$ with the canonical Poisson bracket:

$$\{a,b\} = \Sigma_i \left[\frac{\partial a}{\partial x_i}\frac{\partial b}{\partial v_i} - \frac{\partial a}{\partial v_i}\frac{\partial b}{\partial x_i}\right].$$

The associated group is the group of diffeomorphisms of $\mathbb{R}^6$ that preserve $\{,\}$. The dual space $g^* = \mathcal{D}'(\mathbb{R})$ is the space of distributions on $\mathbb{R}^6$ which we identify with the space of plasma distributions. Using the Lie-Poisson bracket on $g^*$, the flow generated by the Hamiltonian

$$H = \frac{1}{2} \int mv^2 f(x,v)\, dx\, dv + \frac{1}{2} \int E^2\, dx$$

is exactly the Poisson-Vlasov evolution.

The subspace h of g given by affine linear functions of the velocity $\Sigma_i\ a_i(x)v_i + b(x)$ is actually a subalgebra. Working out the Poisson bracket:

$$\{\ \Sigma_i\ a_i^1(x)v_i + b^1(x),\ \Sigma_i\ a_i^2(x)v_i + b^2(x)\}$$

$$= \Sigma_i \left\{\Sigma_j \left[a_j^2 \frac{\partial a_i^1}{\partial x_j} - a_j^1 \frac{\partial a_i^2}{\partial x_j}\right]\right\} v_i + \Sigma_j \left( a_j^2 \frac{\partial b^1}{\partial x_j} - a_j^1 \frac{\partial b^2}{\partial x_j}\right)$$

we see that this subalgebra is isomorphic to the semi-direct product $\mathfrak{X}(\mathbb{R}^3) \times C^\infty(\mathbb{R}^3)$, where $\mathfrak{X}(\mathbb{R}^3)$ is the Lie algebra of vector fields on $\mathbb{R}^3$, acting by differentiation on $\mathbb{C}^\infty(\mathbb{R}^3)$. This Lie algebra is associated with the semidirect product $\mathrm{Diff}(\mathbb{R}^3) \times C^\infty(\mathbb{R}^3)$ of the group of diffeomorphisms of $\mathbb{R}^3$ acting by pullback on the additive group of functions on $\mathbb{R}^3$. This is the group used in the description of compressible fluids [13]. It has also appeared in the "current algebra" approach to quantum field theory [14].

The injection $h \longrightarrow g$ gives us a natural surjection $g^* \longrightarrow h^*$, which is a momentum mapping for the action of the fluid group on the Vlasov plasma manifold. While we have the relation between the Vlasov and fluid Poisson manifolds, complete understanding of the dynamics requires understanding of how the Hamiltonians relate.

Some of the simplifications which were subsystems turn out to be motions on coadjoint orbits. For example the water bag states with fixed area form a coadjoint orbit in the one-dimensional case. The Poisson bracket is just the one usually associated with the KdV equation [15], but the Hamiltonian is different.

## (d) Stability

Another set of interesting mathematical questions may be asked regarding the stability of equilibrium configurations. A good reference for the physics literature earlier than 1960 is [16]. The most detailed work has been done on the 1-dimensional Poisson-Vlasov equations:

$$\frac{\partial f}{\partial t} + v\,\frac{\partial f}{\partial x} + E\,\frac{\partial f}{\partial v} = 0,$$

$$\frac{\partial E}{\partial x} = \int f(x,v)\,dv - 1$$

where the 1 represents an immobile neutralizing background. The distributions $f_0(x,v) = f_0(v)$, $E = 0$, $\int f_0(v)\,dv = 1$ are equilibrium solutions. For the stability analysis we linearize around this fixed point. With Landau [17] we Fourier analyze in x and Laplace transform in t. (This was the beginning of Landau's work with moving poles in particle physics.) One finds that for smooth f the perturbed E-field damps out, an effect now called "Landau damping".

In 1955 N.G. Van Kampen [18] found solutions which are traveling waves for all time. These solutions, however, are not smooth and have the form $g(x,v) = e^{ikx}\left[a\,P\frac{1}{v-V} + b\,\delta(v-V)\right]$. One can connect the two pictures by integrating over the Van Kampen modes. One finds that if the initial $f_0$ decreases with $|v|$,

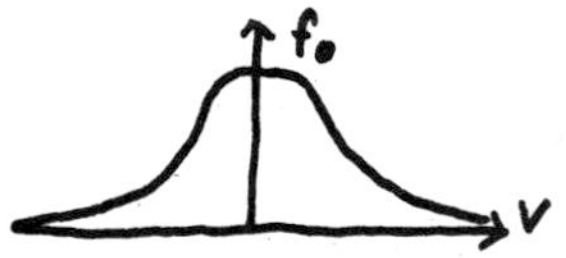


then the solutions to the linear equations remain bounded. If there are bumps in the distribution (beams) then one may find instabilities. In 1960 Penrose [19] found a criterion for stability using the argument principle in locating roots: there is instability when $f_0$ has a local minimum V such that

$$\int \frac{f(v) - f(V)}{(v-V)^2}\,dv > 0.$$

It is a delicate matter to go from neutral linear stability to stability of solutions of the full nonlinear equations. In 1958 Berstein-Greene-Kruskal [20] discovered so called BGK modes. The density of the plasma is constant on level surfaces of the Hamiltonian.

These level curves may look like 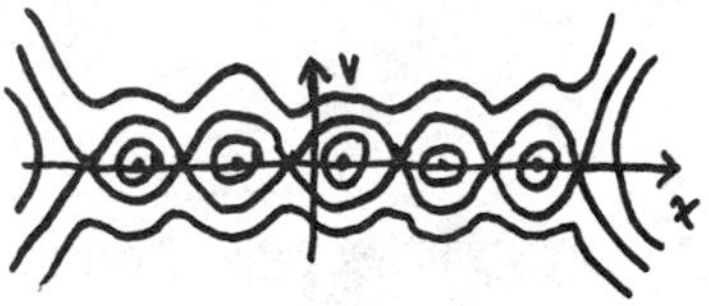


in a reference frame moving with the wave. The density variations exactly produce the electric field necessary to support themselves. These modes are smooth, but as the amplitude gets smaller they may approach the (singular) Van Kampen modes.

It is interesting to try to apply Arnold's stability criterion in this situation. His condition only applies at a regular point of the Poisson structure, since if the dimension of the symplectic leaves goes up when we perturb, there are more directions that we must check for instability. For example, the symplectic leaves for the Lie-Poisson bracket on $sl(2)^*$ look like:  Perturbing from the origin puts us on a 2 dimensional leaf where the dynamics could take us far away. As in the regular point case, we need make no requirements on the Hamiltonian in directions transverse to the symplectic leaves in which the dimension doesn't change.

In the plasma case at hand, the symplectic leaf has infinite codimension. It turns out that we can get by with restricting to only a codimension 2 surface, defined by level sets of two functions. Anything of the form $\int G(f(x,v))\, dx\, dv$ is constant on coadjoint orbits. For the first function we may choose the total number of particles $\int f\, dx\, dv$. With this choice, the 2nd function is determined by the equilibrium point we are interested in. For the case of a Maxwellian distribution $\alpha e^{-v^2}$ the function turns out to be the entropy: $\int f \log f\, dx\, dv$. This is related to the fact that the distribution function which maximizes the entropy for given total particle number is the Gaussian. This line of thinking has only begun, and there is lots left to be done.

## Bibliography

[1] W. Crookes, Phil. Trans. 1, (1879), 135.

[2] See e.g. Jackson, J.D., Classical Electrodynamics, 2nd edition, John Wiley and Sons Inc., New York (1975).

[3] Some general references on plasma physics are:

**Introductory:**

F. Chen, Introduction to Plasma Physics, Plenum, New York (1974).

**More Advanced:**

P.C. Clemmow and J.P. Dougherty, Electrodynamics of Particles and Plasmas, Addison-Wesley, Reading, Mass. (1969).
N. Krall and A. Trivelpiece, Principles of Plasma Physics, McGraw-Hill, New York, (1973).
G. Schmidt, Physics of High Temperature Plasmas, Academic Press, New York, (1979).
S. Ichimaru, Basic Principles of Plasma Physics, W.A. Benjamin, Inc., Reading, Mass. (1973).
R.C. Davidson, Methods in Nonlinear Plasma Theory, Academic Press, New York, (1972).

[4] See e.g. S. Wollman, Comm. Pure Appl. Math 33 (1980) 173-197.

[5] F. Hohl and M.R. Feix, Astrophys. J. 147, (1967), 1164.

H.L. Berk, C.E. Nielson and K.V. Roberts, Phys. Fluids 13 (1970), 980.

[6] N.J. Zabusky, Ann. N.Y. Acad. Sci. 373 (1981), 160-170.

[7] V. Arnold, Annales de l'Institut Fourier 16 (1966), 319-361.

[8] A. Weinstein, "The local structure of Poisson manifolds", J. Diff. Geom., to appear (1983).

S. Lie, Theorie der Transformationgruppen, Zweiter Abschnitt, Teubner, Leipzig (1890).

[9] F.A. Berezin, Funct. Anal. Appl 1 (1967), 91.

R. Hermann, Toda Lattices, Cosymplectic Manifolds, Bäcklund Transformations, and Kinks, Part A, Math. Sci. Press, Brookline (1977).

A. Lichnerowicz, J. Diff. Geom. 12 (1977), 253.

[10] P.J. Morrison, Phys. Lett. 80A, (1980), 383.

[11] A Weinstein and P.J. Morrison, Phys. Lett. 86A, (1981), 235.

[12] J. Marsden and A. Weinstein, "The Hamiltonian structure of the Maxwell-Vlasov equations", Physica D 4, (1982), 394.

[13] J.E. Marsden, T. Ratiu, and A. Weinstein, "Semi-direct Products and Reduction in Mechanics", Trans. Amer. Math. Soc., to appear, (1983).

[14] G.A. Goldin, R. Menikoff and D. J. Sharp, J. Math. Phys. 21 (1980), 650.

[15] C.S. Gardner, J. Math. Phys. 12 (1971), 1548.

V.E. Zakharov and L.D. Faddeev, Funct. Anal. Appl. 5 (1971), 280.

[16] J.D. Jackson, J. Nuclear Energy C 1 (1960), 171.

[17] L.D. Landau, J. Phys. U.S.S.R. 10, (1946), 25.

[18] N.G. Van Kampen, Physica 21 (1955), 949.

[19] O. Penrose, Phys. Fluids 3 (1960), 258.

[20] I.B. Bernstein, J.M. Greene, M.D. Kruskal, Phys. Rev. 108 (1957), 546.